PRINCIPLES OF RADIOGRAPHY FOR TECHNOLOGISTS

PRINCIPLES OF RADIOGRAPHY FOR TECHNOLOGISTS

Perry Sprawls, PhD
Department of Radiology
Emory University School of Medicine
Atlanta, Georgia

AN ASPEN PUBLICATION®
Aspen Publishers, Inc.
Rockville, Maryland
1990

Library of Congress Cataloging-in-Publication Data

Sprawls, Perry.
Principles of radiography for technologists/Perry Sprawls.
p. cm.
"An Aspen publication."
ISBN: 0-8342-0088-0
1. Radiography, Medical. I. Title
[DNLM: 1. Technology, Radiologic. WN 160 S767p]
RC78.S685 1989
616.07'572—dc20
DNLM/DLC   89-17719
for Library of Congress   CIP

Aspen Publishers, Inc., grants permission for photocopying for limited personal or
internal use. This consent does not extend to other kinds of copying, such as
copying for general distribution, for advertising or promotional purposes, for
creating new collective works, or for resale. For information, address Aspen
Publishers, Inc., Permissions Department, 1600 Research, Rockville, Maryland
20850.

Much of the artwork in this text was previously
published in Physical Principles of Medical Imaging by P. Sprawls Jr., Aspen
Publishers, Inc., © 1987.

The authors have made every effort to ensure the accuracy of the information herein,
particularly with regard to drug selection and dose. However, appropriate information sources should be consulted, especially for new or unfamiliar drugs or
procedures. It is the responsibility of every practitioner to evaluate the appropriate-
ness of a particular opinion in the context of actual clinical situations and with due
consideration to new developments. Authors, editors, and the publisher cannot be
held responsible for any typographical or other errors found in this book.

Editorial Services: Jane Coyle Garwood

Library to Congress Catalog Card Number: 89-17719
ISBN: 0-8342-0088-0

Printed in the United States of America

1 2 3 4 5

Table of Contents

Preface

Radiography continues to be one of the medical profession's most valuable diagnostic techniques. However, its effectiveness depends on the quality of the radiographic images, which in turn relies on the education, experience, and motivation of the radiographers and technologists.

Principles of Radiography for Technologists provides the student technologist with a knowledge of the basic principles of radiography and how they are applied to obtain appropriate image quality. The material in this book is useful at all levels of the educational process. The new student in radiography will find it a comprehensive introduction to the process of x-ray imaging. Advanced students will find it useful in preparing for examinations, and staff technologists can use it as a general reference and for continuing education activities.

Principles of Radiography for Technologists is a companion book to the *Physical Principles of Medical Imaging*, which is used as a text by residents in radiology. It is hoped these books provide for better communications and a more effective working relationship within the x-ray team.

Acknowledgments

The preparation of this book has been aided by the significant contributions of many individuals, whom I gratefully acknowledge.

Margaret Nix typed and edited the manuscript and coordinated its production. Dr. Jack E. Peterson has provided much valuable advice and editorial support. Jake D. Paulk, Alan Foust, and Lee Burns produced the illustrations. My wife Charlotte has graciously provided editorial assistance.

The support and guidance of Dr. William J. Casarella, Chairman, Department of Radiology, have been significant factors in the development of the educational programs that are the source of material for this book.

Chapter 1

The Radiograph

Radiography is the process of producing x-ray images that are recorded on film. These images are known as *radiographs*. All x-ray images are not radiographs. Two examples are fluoroscopic images and computed tomography (CT) images.

One of the major concerns in radiography is to produce good radiographs. A good radiograph is one that gives the physician an adequate view of the anatomic structures and changes in the tissue so that an accurate diagnosis can be made. Different clinical applications, such as chest radiography, mammography, and angiography, require radiographic images with somewhat different characteristics. The radiographic procedure and technique must be adjusted by the radiographer so that it produces a radiograph that meets the needs of the different clinical procedures. This book is about producing good radiographs.

In this chapter we will begin the process of producing good radiographs by first learning more about the radiograph itself. Radiographs have certain characteristics that together will determine if they are good or not. In order for a radiograph to have good image quality, all of its characteristics must be satisfactory. Remember, a chain is no stronger than its weakest link. If a radiograph has one unsatisfactory characteristic, it will not be good even if all of the other characteristics are satisfactory. Let's now look at a radiograph and find out what are the characteristics that must be considered. Figure 1-1 is a typical radiograph. If an actual radiograph is available to you, it will be helpful to look at it during our discussion.

DENSITY

When we look at a typical radiograph, we see some areas that are very bright, some areas that are very dark, and some areas that are in between, with different shades of gray. The darkness in a radiograph is known as film *density*. Place your finger on one of the light areas in the radiograph. That is an area that can be

1

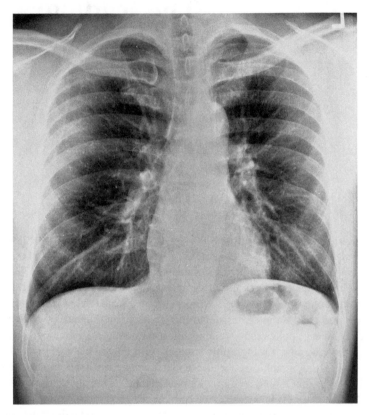

Figure 1-1 A typical radiograph showing areas with different film densities.

described as having a low density. The very dark areas have a high density, and the gray areas can be described as having a medium density.

Radiographs are viewed on a lighted view box. It is the density in the film that absorbs the light and keeps it from shining through. On the other hand, the areas with low density allow most of the view box light to shine through, causing them to be light or bright. The amount of density at any point in a radiograph can be expressed with a number value, but we will discuss that in a later chapter.

Density is an important characteristic of a radiograph. If it is too dense (dark), the light cannot shine through to show the image. If the density is too low (light), it usually indicates that there has not been sufficient exposure to the film to form a good image.

Film density is produced by exposure to the film. A dense (dark) area is one that has received more exposure than a less dense (light) area. Look at the radiograph and locate the areas that appear to have received the highest x-ray exposure. Now locate some of the areas that have received the lowest exposure.

CONTRAST

Another important characteristic of a radiographic image is *contrast*. Let's begin by considering the general meaning of the word "contrast." Contrast means difference. We sometimes speak of the contrast between people who are very different in some way. If two colors are very different, we might say that there is much contrast between them. Contrast in a radiograph is the difference in density from one point to another. When you look at a radiograph and see the different densities, you are seeing contrast. If there were no contrast or differences in density, there would be no image. It is probably fair to say that contrast is the most important characteristic of a radiographic image.

If an image is described as having high contrast, this means that it contains both high density (dark) and low density (light) areas. A low-contrast radiograph is one in which there are relatively small differences in density throughout the image. Before we can show the relationship between image contrast and image quality, we must consider two types of contrast found in images.

Area Contrast

Look at Figure 1-1 and notice that there are some relatively large areas that are dark (high density) and some areas that are rather light (low density). These areas often correspond to the major organs or parts of the skeletal system. For example, notice that the lung areas are much more dense than the area of the heart (the mediastinum). In other words, there is contrast (a difference in density) between these two major areas. This is *area contrast*. Too much area contrast is an undesirable characteristic of an image. The problem with high area contrast is that some areas are too dense and dark and other areas are not dense enough to show the anatomy. Figure 1-2 shows three images. The first image (1-2A) has too much area contrast compared to the last image (1-2C). Notice that you can see many more parts of the body in the image with the least area contrast.

Certain parts of the human body, such as the chest, often produce x-ray images with too much area contrast. Specific techniques must be used to reduce the contrast in the images.

Object Contrast

A good radiograph is an image in which many individual objects within the human body are visible. These objects include the normal anatomical structures, such as bones and blood vessels, as well as abnormal objects like masses or fractures that are caused by disease and injury. Any specific object will be visible in an image only if it produces contrast with respect to its background. Consider

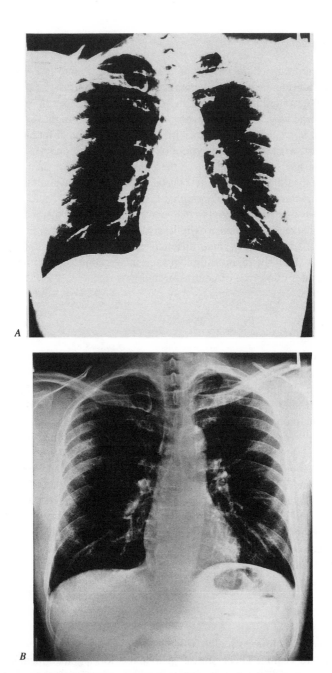

A

B

Figure 1-2 Three images with different amounts of area contrast. The one on the next page (1-2C) with the least area contrast has better object visibility in both the light and dark areas.

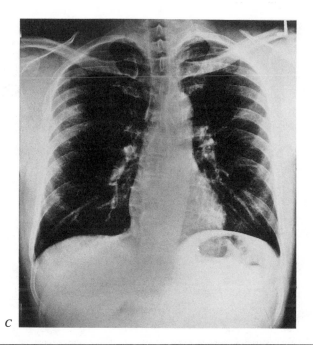

C

Figure 1-3. Here we see a simple round object imaged with different amounts of contrast. Notice how it becomes more visible as the contrast between the object and the background increases. A general goal of a radiographic procedure is to produce an image in which the various objects in the body produce enough contrast so that they are readily visible.

We have now observed two basic types of contrast in a radiograph. *Object contrast* is a desirable characteristic that makes objects in the body visible. But if an image has too much area contrast, it will have the effect of reducing the object contrast and visibility of objects within the very dark or very light areas. The specific cause of this will be discussed later. Object contrast is desirable because it is what makes individual objects in the body visible. However, too much area contrast is undesirable because it can reduce object contrast and object visibility in certain areas.

DETAIL

The human body contains objects, both normal and abnormal, with a wide range of sizes. There are relatively large objects like vertebrae and ribs and small objects such as trabecular patterns within bone, fractures, and small calcifications. The

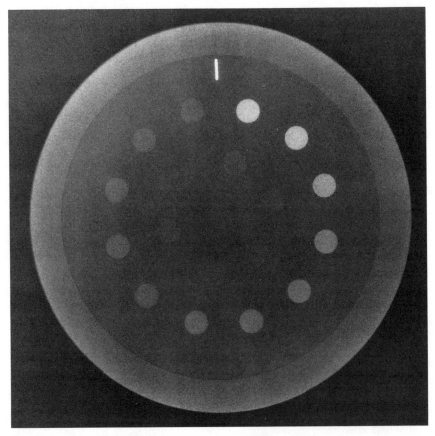

Figure 1-3 Objects become more visible when the contrast between the object and its background is increased.

small objects and anatomical structures are the *detail* in the body. An important characteristic of a radiograph is the visibility of small objects (anatomical detail).

Look at the two images in Figure 1-4. You will notice that there is much more anatomical detail visible in the image on the right (B). The image on the left (A) does not show as much visibility of detail because it is blurred more than the image on the right.

Blurring is what reduces visibility of detail in all kinds of images. When our vision is blurred, we cannot see details like small print. We sometimes see photographs that do not show much detail because the image was blurred by someone's moving or the camera's being out of focus.

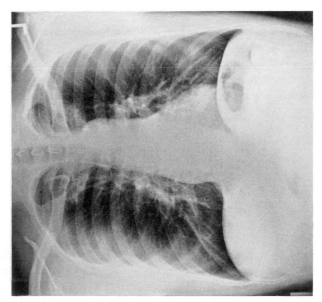

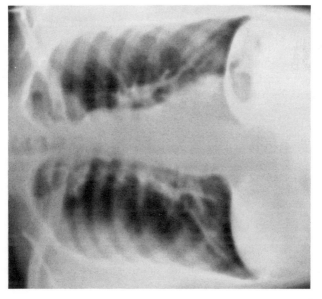

B

A

Figure 1-4 The image on the right (*B*) shows better visibility of anatomical detail.

In radiography there are three specific things (motion is one of them) that blur the images to some extent. An image with a relatively large amount of blur will have less visibility of detail.

Some radiographic procedures, such as mammography, require much more visibility of detail than others. Therefore, specific steps must be taken to reduce blur to an acceptably low value.

NOISE

Look closely at the two images shown in Figure 1-5. How would you describe the difference? Notice that the image on the left has an appearance that might be described as grainy or textured. The appropriate name for this is *image noise*, although you might find it also referred to as *image mottle*.

The reason why it is named noise is because it interferes with our ability to see some objects just like audio (sound) noise interferes with our ability to hear and understand softly spoken words. Visual noise in radiographs interferes with the visibility of objects that have relatively low contrast.

All radiographs have some noise, but it is usually less than the noise in most other forms of medical images, such as fluoroscopy and computed tomography.

ARTIFACTS

Occasionally we will see something in an image that was not produced by the patient's body. These undesirable things are called *artifacts*. Figure 1-6 shows a rather common artifact. Artifacts come from a variety of sources, including objects outside of the patient's body, scratches on the film, and static electricity.

DISTORTION

A radiograph generally gives us a slightly distorted view of the human body. Figure 1-7 illustrates three forms of radiographic distortion.

Shape

In many instances a radiograph will not show the true shape of an object. This is usually related to the orientation of the object with respect to the direction of the x-ray beam.

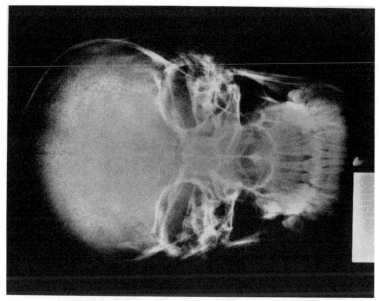

B

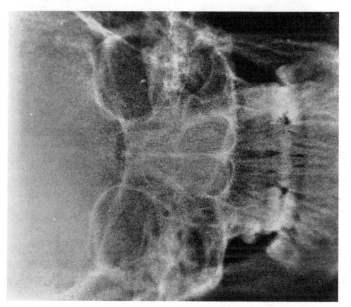

A

Figure 1-5 The image on the left (A) has more noise than the one on the right (B).

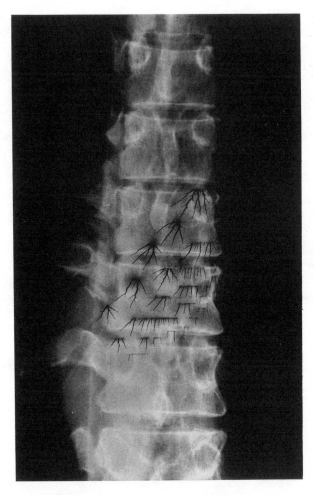

Figure 1-6 Radiographic artifact.

Size

A radiograph does not usually show the true size of an object. All objects in the human body are magnified to some extent in the final image. The amount of magnification depends on the location of the object. Therefore, objects at different locations within the patient's body will be magnified by different amounts and will have different amounts of size distortion.

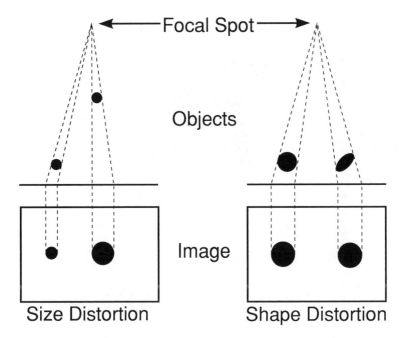

Figure 1-7 Examples of radiographic distortion.

Position

A radiograph can also distort the relative position or distance between objects within the body. This is usually associated with the magnification effect. Consider two pairs of objects that are located different distances from the receptor. In both pairs the objects are actually located the same distance apart. However, in the image the distance between the objects appears to be quite different because one pair is magnified more than the other. The orientation of the objects in relationship to the direction of the x-ray beam has a significant effect on the amount of distortion.

SUMMARY

We will use Figure 1-8 to bring together and summarize the different charac-teristics of a radiograph. All five of these characteristics (contrast, detail, noise, artifacts, and distortion) contribute to the overall quality of the image. It is

IMAGE QUALITY

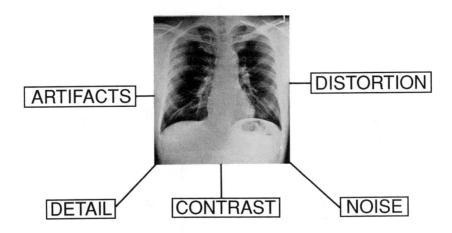

Figure 1-8 The five characteristics of a radiograph that determine its quality.

important to recognize that image quality is not determined by one single characteristic but by the combination of the five characteristics taken together.

In future chapters we will consider each of these characteristics in much more detail and show how they are affected by the various parts of the radiographic equipment system or by the selection of technical factors used by the radiographer.

The radiation exposure to a patient is an issue that must be considered in radiography. It should be kept as low as is practical. We will soon discover that often it is necessary to make compromises between image quality and patient exposure. There are several situations in which image quality can be improved by using more exposure to the patient. In setting up radiographic procedures and selecting technique factors, we need to achieve a proper balance between image quality and patient exposure.

Study Activities

Name the five basic characteristics of a radiographic image that determine its quality.

While looking at a radiograph, locate the areas that appear to have the highest film density. Now locate the areas with the lowest film density.

While looking at the same radiograph, locate the area that you think received the highest exposure.

Describe the relationship between film exposure and film density.

While looking at a radiograph, locate two areas that appear to have the most contrast between them.

While looking at a radiograph, locate the objects that appear to have the highest contrast with respect to their background.

While looking at a radiograph, locate some objects that have a low contrast in relationship to their background.

Describe how the contrast of an object affects its visibility.

Find some radiographs in which a high *area contrast* reduces the object contrast and visibility in both the light and dark areas.

While looking at a radiograph, locate and name several small objects (anatomical structures).

While looking at the same radiograph, squint your eyes so that your vision is blurred. Describe the effect of blurring on the visibility of detail.

Find some radiographs in which motion has blurred the visibility of anatomical detail.

While looking at a radiograph, find the visual noise. Try to find some low contrast objects whose visibility is affected by the noise.

Look at a fluoroscopic image and compare the noise to what you see in a radiograph.

Identify and make a list of artifacts that you have seen in radiographs.

Explain how the different types of distortion can be produced in radiography.

Chapter 2

Radiographic Equipment

A radiographic system ("x-ray machine") is made up of several major components that work together to produce a radiograph of the patient's body. Each of the components has certain characteristics that must be considered when using the equipment. All radiographic equipment is not the same. The different types of x-ray examinations require equipment with different characteristics in order to produce images with the appropriate quality.

When performing an x-ray examination, the radiographer should make sure that the equipment is appropriate for the specific type of examination. For example, a mammogram requires equipment with characteristics that are very different from the equipment for chest radiographs.

A radiographic system also has several technical factors that must be selected and set by the radiographer in order to produce a good image. The factors used for a specific examination are usually known as the *technique*. A radiographer will usually know the general technique for many types of examinations but will also use technique charts to provide additional information.

A technique chart shows the appropriate exposure factors to use to radiograph different parts of the body. It also shows how to adjust the exposure for patients of different sizes. It should be remembered that all x-ray equipment is different. Each machine should have its individual technique chart to ensure the best image quality possible.

In this chapter we will begin to look at the radiographic equipment and identify the major components of the system. As we consider each component, we will also identify the specific characteristics of that component that will have an effect on image quality, patient exposure, or some other important factor.

As you study this chapter I suggest that you take a look at some radiographic rooms, if possible, and locate the different components as they are discussed. Later chapters will give much more detailed information on the various parts of the equipment and how they are to be used.

Figure 2-1 shows some of the major components of a radiographic system.

15

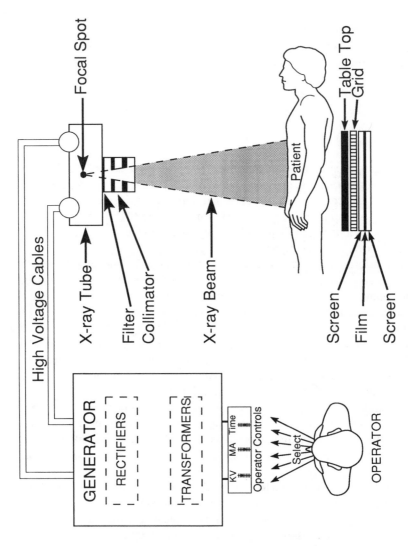

Figure 2-1 Some of the basic components of a radiographic system.

THE X-RAY TUBE

The *x-ray tube* produces the x-ray beam. In some ways it can be compared to a light bulb. When we provide a light bulb with electrical energy by turning it on, it produces light and heat. When we turn on an x-ray tube by pushing the exposure switch, we provide it with electrical energy. The x-ray tube converts the electrical energy into x-radiation and heat. The heat is an undesirable by-product of the x-ray production process. We must be careful not to overheat the x-ray tube and damage it.

When you look at an x-ray tube, you see a metal cylinder attached to two large electrical cables. This is the x-ray tube *housing*. The parts of the x-ray tube that produce the radiation are contained in a glass tube, called the insert, located inside the metal housing.

The component within the insert that actually produces the x-ray beam is the *anode*. The radiation comes from a small spot on the surface of the anode, known as the *focal spot*. The size of the focal spot is a very important characteristic to be considered when using an x-ray tube. X-ray tubes are manufactured with a variety of focal-spot sizes, ranging from approximately 0.1 mm up to 2 mm. Most tubes have two focal spots, small and large, from which you can make a selection when setting up a technique. On the x-ray tube housing you will find a small metal tag showing the size of the focal spot. You should be familiar with the focal spot size for the equipment you are using.

In general, small focal spots are selected to give better visibility of detail. However, the larger focal spot is sometimes required in order for the tube to be able to tolerate the heat that is produced during the exposure.

Another important characteristic of any specific x-ray tube is the amount of heat it can contain without producing damage. This is described by the tube's *rating chart*. The rating chart is supplied by the tube manufacturer and is usually displayed on the wall in the control booth. Later we will learn how to use rating charts, but for now you should locate the charts that apply to any equipment you are using.

BEAM-LIMITING DEVICES

There must always be some device attached to the x-ray tube that gives the x-ray beam a specific size and shape. For most radiographic applications this device is a *collimator*. Collimators usually produce square or rectangular beam areas or *fields of view* (FOV). Control knobs on the collimator are used to adjust the length and width of the FOV.

Some collimators have an automatic feature that prevents the FOV from being larger than the film. This is known as a *positive beam-limiting (PBL) device*.

The collimator contains a light source that projects a beam of light along the same pathway as the x-ray beam. The light is turned on temporarily to position the x-ray beam with respect to the patient's body and to visualize its size and shape.

Some special-purpose radiographic equipment might use devices other than collimators for limiting the x-ray beam. These devices are cones or diaphragms, which produce a fixed FOV.

FILTER

All radiographic equipment must filter the x-ray beam before it reaches the patient's body. The *filter* is a thin sheet of metal (usually aluminum), typically located between the x-ray tube and the collimator. The purpose of the filter is to absorb some of the excess x-ray beam that is not very good for producing an image but would produce unnecessary exposure to the patient. General-purpose radiographic equipment should have at least 2.5 mm of filtration. The amount of filtration is not one of the changeable technique factors.

GRID

After the x-ray beam passes through the patient's body, it reaches the receptor, which converts the invisible x-ray beam into an image on the film. However, it passes through a *grid* before it reaches the receptor (which will be discussed below). The purpose of the grid is to absorb some of the scattered radiation that is produced within the patient's body as a result of the passage of the x-ray beam. Since this scattered radiation can expose the film just like the x-ray beam, it must be removed in order to improve the contrast characteristics of the image. In a later chapter we will learn how a grid performs this function, but for now we will identify those characteristics that affect a grid's performance.

The grid looks like a sheet of metal that has the same size as the receptor. The grid actually is made up of many small lead strips which we cannot see. The lead strips are arranged to absorb the scattered radiation as described in more detail in a later chapter. The grid is located at the entrance surface to the receptor. In many systems the grid is mounted in a device that moves it during the exposure. This device is generally known as the Bucky mechanism. The movement keeps the grid from producing small, closely spaced lines (grid lines) in the image, which may appear if the grid is not moved.

You should go and inspect the grids on several radiographic systems if they are accessible. Each grid will have a label indicating its specific characteristics.

Grid Ratio

The primary characteristic that makes grids differ from one another is the *grid ratio*. Grid ratio is related to the dimensions of the small lead strips within the grid. Grid ratio values are generally in the range of 5 to 1 up to 16 to 1. High-ratio grids usually produce better images, but they also increase patient exposure and the heating of the x-ray tube and make positioning more difficult. The selection of a grid for a specific clinical procedure is a compromise among these factors.

Focal Distance

There are two types of grids: focused and nonfocused. If a grid is of the focused type, it will have a focal distance indicated on the label. This is the distance that must be present between the grid and the focal spot of the x-ray tube.

Tube Side

Focused grids must always be positioned with a specific side of the grid closest to the x-ray tube. A label will indicate which is the tube side.

RECEPTOR

The *receptor* is the device that receives the x-ray beam and produces the image on the film. Most radiographic receptors are in the form of cassettes. The cassette contains two major components: *intensifying screens* and *film*. Most cassettes contain two intensifying screens, with the film sandwiched in between them. The function of the intensifying screens is to absorb the x-ray beam and convert it into light, which then exposes the film. This process makes it possible to create an image with much less x-ray exposure, compared to letting the x-ray expose the film directly. Intensifying screens have two major characteristics that must be considered when selecting screens for a specific clinical application. One is their *sensitivity*, or the amount of radiation exposure required to produce an image. This is sometimes referred to as the *speed* characteristic. The other major factor is *the amount of image detail* a specific intensifying screen can produce. Unfortunately, screens that produce maximum image detail require more exposure (ie, are less sensitive) than screens that produce less detail. This is a compromise that must be made when selecting screens.

Film

Radiographic film consists of an almost clear base material with a thin emulsion coated on both sides. The emulsion contains the chemicals that convert exposure into a visible image. The image is formed in the emulsion. Film has two major characteristics that must be considered when selecting film for a specific clinical application and when selecting technique factors. One factor is its sensitivity (speed), and the other factor is the film's contrast characteristics. Some film will produce a "more contrasty" image than other film. We will explore that subject in a later chapter.

THE GENERATOR

If we follow the two large electrical cables running from the x-ray tube, they will lead us to the part of the system known as the *generator*. The generator consists of several major components, some of which we will introduce here.

The major function of the generator is to supply the electrical energy to the x-ray tube, which it uses to produce the x-ray beam. The generator does not actually generate or produce the electrical energy. It receives energy from the power company and changes it into a form the x-ray tube can use. In principle, the generator "repackages" the electrical energy before sending it to the x-ray tube. The generator also provides us with the ability to control certain characteristics of the electrical energy being supplied to the tube. These three important characteristics are (1) the kilovoltage—KV, (2) the tube current—MA, and (3) the exposure time.

We will soon find that these factors have a significant effect on the characteristics of the x-ray beam, which will in turn have an effect on both image quality and patient exposure.

Transformer

The *transformer* is a rather large device that is usually located in either a corner of the x-ray room, behind panels, or in an adjacent equipment room. It is the device that the large cables from the x-ray tube connect to. The transformer plays a role in "repackaging" the electricity.

Electrical Equipment Cabinets

A radiographic system might have one or more metal cabinets containing various parts of the generator. These cabinets generally do not contain any components or controls that are to be adjusted by the radiographer.

Control Console

The *control console* is the part of the generator that we use to set some of the technique factors and also to observe various meters and indicators.

DISTANCE

Figure 2-2 shows some of the distances important in a radiographic system. Distances are a factor that must be considered by the radiographer when setting up an examination, because they have an effect on the characteristics and quality of the image.

Focal Spot-Receptor Distance

One factor that is adjusted by the radiographer is the distance between the focal spot and the receptor. This is often designated as the *focal-receptor distance* (FRD). This distance is usually adjusted by moving the x-ray tube, and a scale or indicator is usually provided for reading the distance. A long FRD usually

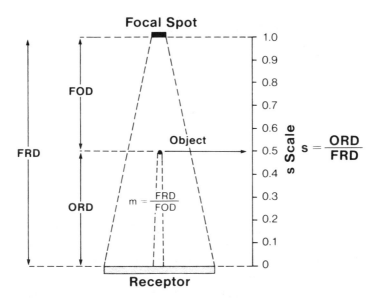

Figure 2-2 Distances that have an effect on image characteristics and quality.

produces an image with less distortion and focal spot blurring but has the disadvantage of requiring more exposure from the x-ray tube. There are some common FRD values associated with specific examination types. Examples are

- Chest: 72 inches
- General radiography: 36 to 45 inches
- Mammography: 20 to 30 inches

Object-Receptor Distance

The distance between the various objects in the patient's body and the receptor has an effect on image distortion and detail, as we will see in a later chapter. In many radiographic examinations this distance is relatively small because the patient's body is located as close as possible to the receptor. Image detail is affected by the object-receptor distance (ORD) in proportion to the FRD. Therefore, it is often helpful to represent the ORD as a fraction of the FRD, as shown in Figure 2-2. This is known as the s scale, which always runs from a value of zero at the receptor to a value of 1 at the focal spot. The object location along the s scale is

$$\text{Object location (s)} = \frac{\text{ORD}}{\text{FRD}}$$

There are examinations in which the distance between the body and the receptor is increased for the purpose of producing image magnification. The magnification factor is the ratio of two distances:

$$\text{Magnification} = \frac{\text{FRD}}{\text{FOD}}$$

Focal-object distance (FOD) is the ORD subtracted from the FRD. If the FRD is not changed, magnification will increase if the ORD is increased. When the ORD is increased, the FOD decreases, and this increases the magnification ratio.

SUMMARY

The radiographic equipment system consists of many different components. The individual components have specific characteristics that the radiographer must be familiar with when selecting equipment for a specific type of examination and when setting the technique factors.

We will consider the characteristics of each of the components in much more detail in later chapters.

Study Activities

Make a list of the different components of a radiographic system.

Take this list to several radiographic rooms and locate each component.

Locate the label on the x-ray tubes and determine the size of the focal spots.

Examine the control console and determine how you select between the small and large focal spots.

Locate the rating charts for the x-ray tube that you use.

Determine the amount of filtration in the radiographic equipment. (Your supervisor or physicist can provide this information.)

Inspect the grid to determine its ratio and focal distance.

Examine the cassettes and determine what type of intensifying screens are used.

Determine what type of film is used.

Determine the speed number for the cassettes.

Locate the high-voltage transformer and the cables that connect it to the x-ray tube.

Examine the equipment and determine how the focal spot-receptor distance (FRD) is measured.

Determine whether the equipment is operated with single-phase or three-phase electricity.

Chapter 3

The X-Ray Beam

The *x-ray beam* is what passes through the patient's body and forms the image that is recorded on the film. The quality of the image often depends on the characteristics of the x-ray beam. Therefore, when setting up a technique, we must adjust the x-ray beam to match the needs of the examination.

In this chapter we will learn more about what's in an x-ray beam and its important characteristics that must be considered when setting up a radiographic procedure.

RADIATION AND ENERGY

The x-ray beam is a form of *radiation*. Radiation is energy that is moving through space. In everyday life we encounter several different types of radiation. The most common, and the one that we can see, is light. Radio and television signals that bring programs into our homes are another form of radiation. Microwave ovens use a form of radiation to produce heat in our foods. All of the radiations mentioned up to this point belong to the same general family. The name of the family is *electromagnetic* radiation. As with any family, the individual members have some characteristics in common and some characteristics that make them different from one another.

Let's now look at some of the characteristics of an x-ray beam and see how it compares to the more familiar form of radiation, light. At first we encounter a small problem because the x-ray beam is invisible. However, by using our imagination and Figure 3-1, we can get a good picture of an x-ray beam in our minds.

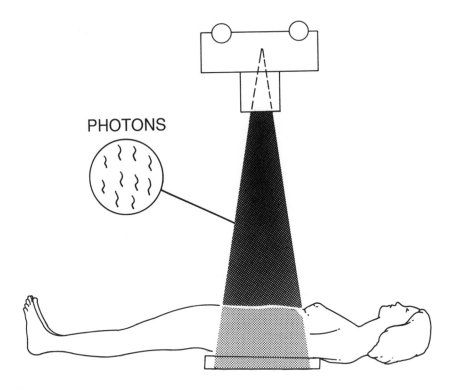

Figure 3-1 An x-ray beam, showing its photon structure.

Photons

If we could look at an x-ray beam through an imaginary super-duper micro-scope, we would see that it is made up of many small individual units of energy named *photons*. A photon contains nothing but energy. It has no mass or weight. An x-ray beam is actually a shower of many, many photons. Each time we push the exposure switch of the radiographic system, we are turning on the shower.

Photons fly through space at a very high speed (velocity). In fact, all photons of the electromagnetic radiation family travel with the same velocity. Since light is the most familiar member of that family, we can describe x-ray photons, and the x-ray beam, as traveling with the speed of light. X-ray photons have very short lifetimes. As soon as they are created in the x-ray tube, they go flying away at the speed of light. They are then very quickly absorbed in either the patient's body, the intensifying screens of the receptor, or any other material they might encounter. Since all photons travel at the same velocity there's nothing we can do to speed them up or slow them down.

The most important characteristic of an individual x-ray photon is the amount of energy it contains. The reason why this is important is that photons with certain energies are much better for producing certain types of x-ray images than other photons. This is a factor that must be considered when setting up a technique, because we do have some control over the energy of the photons.

The unit that is used to express the energy of individual photons is the *kiloelectron volt* (keV). The range of photon energies found in x-ray beams used for radiography is generally from about 20 to 120 keV, but this depends on the technique factors being used.

X-Ray Spectrum

Every x-ray beam contains photons with many different energies. The content of an x-ray beam with respect to different photon energies is referred to as its *spectrum*. We can draw a picture of an x-ray beam's spectrum, as in Figure 3-2. This is a simple graph showing the relative number of photons at different energies.

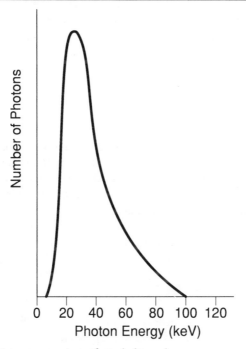

Figure 3-2 The photon energy spectrum of a typical x-ray beam.

Look at the spectrum in Figure 3-2 and answer the following questions:

1. What is the maximum photon energy in this beam?
2. What is the minimum photon energy?
3. What photon energy has the greatest number of photons in the beam?

Later we will consider how to change the spectrum, or photon energy content, of our x-ray beams.

X-RAY BEAM QUALITY

Quality is a characteristic of an x-ray beam that relates to its photon energy content or spectrum. An x-ray beam that contains many low-energy photons is described as being ''soft,'' whereas a beam that has most of its photons at the higher energies is referred to as a ''hard'' beam.

The primary significance of x-ray beam spectrum and quality is that these are the characteristics determining the penetrating ability of the radiation. Low-energy (soft) photons are generally not very good penetrators of materials such as human tissue. High-energy (hard) photons are generally better penetrators. The quality of an x-ray beam is one of the factors that has an effect on image contrast.

In Chapters 7 and 10 we will find that there are several factors that have an effect on the x-ray beam spectrum and its quality. However, the KV is the only one that is a technique factor and that can be changed by the radiographer.

X-RAY BEAM QUANTITY

Another important characteristic of an x-ray beam is the amount or *quantity* of radiation it contains. In the next chapter we will find that there are several different quantities used to express the amount of radiation. The most common quantity is exposure, whose measure expresses the concentration of radiation at some specific point. Exposure is expressed in roentgens.

The amount of exposure produced by an x-ray beam is controlled by the three technique factors—KV, MA, and exposure time.

INVERSE-SQUARE EFFECT

A characteristic of an x-ray beam or any other type of radiation emitted from a relatively small source is that it is constantly spreading out (diverging) as it moves away from the source, as shown in Figure 3-3. At any point along the beam, the

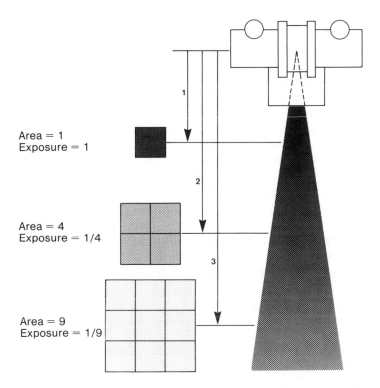

Area = 1
Exposure = 1

Area = 4
Exposure = 1/4

Area = 9
Exposure = 1/9

Figure 3-3 Inverse-square effect.

width of the area covered is proportional to the distance from the source. At a distance of 1 m, the cross-section beam area is one unit wide. It progresses to a width of two units at a distance of 2 m. At 3 m the area is three units by three units. Therefore, the area covered by our x-ray beam is increasing in proportion to the square of the distance from the source. For example:

Distance	Area
1	$1 \times 1 = 1$
2	$2 \times 2 = 4$
3	$3 \times 3 = 9$

First, let us consider the amount of radiation passing through the three areas. We are assuming that none of the radiation is absorbed or removed from the beam before it reaches the third area. All radiation that passes through the first area will also pass through the second and third areas. In other words, the total amount of radiation is the same through all areas and does not change with distance from the source.

Now let us consider the exposure, which is the concentration of radiation through the three areas. In the first area, all radiation is concentrated in a one-unit area. At a distance of 2 m from the source, the radiation is spread over a four-unit (two by two)-square area and continues to spread to cover a nine-unit (three by three)-square area at a distance of 3 m. Since the same total amount of radiation is being distributed over larger areas as distance increases, the radiation is obviously becoming less concentrated, and the exposure is decreased.

What we have observed here is the fact that as radiation moves away from its source, the total amount of radiation does not change, but its concentration (exposure) decreases. At any given distance, the exposure is inversely proportional to the area covered by the beam. This means that as the area increases, the exposure decreases because the radiation is now spread over a larger area. The area covered by the beam is directly proportional to the *square* of the distance from the source. For example, as the distance increases from 1 m to 2 m, the area increases from $1 \times 1 = 1$ to $2 \times 2 = 4$. We can conclude that the exposure (concentration) of radiation is inversely related to the square of the distance from the source. This is commonly known as the *inverse-square law*.

In radiography we must take this into account any time we change the distance between the focal spot and the receptor (FRD). If the distance is increased, the exposure to the receptor will be decreased. In this case it will be necessary to change the technique factors in order to produce more radiation in the tube, to compensate for the increased distance.

Study Activities

Name several types of radiation that are members of the electromagnetic radiation family.

Explain how x-radiation and light are alike.

Explain how x-radiation and light are different.

Name the one most important characteristic of a photon.

Draw a spectrum for an x-ray beam that has a minimum photon energy of 20 keV and a maximum photon energy of 85 keV.

Label the areas of the spectrum that represent soft and hard radiation.

Name the technique factor you can use to change the energy spectrum or quality of an x-ray beam.

Name the technique factors you can use to change the quantity of radiation or exposure produced by an x-ray beam.

Determine the exposure at a distance of 1 m from the focal spot if the exposure at 2 m is 2 roentgens.

Chapter 4

Radiation Quantities and Units

There are several different quantities used to express amounts of radiation. The reason why there are different quantities is because each one expresses a different characteristic of the radiation. It's somewhat like expressing the size of a person. We cannot do it with a single quantity. We must use several quantities, such as height, weight, shoe size, etc. Each of these quantities is related to a different characteristic of a person.

The radiation quantities we are about to consider fall into two categories, as shown in Figure 4-1. Three of the quantities—exposure, dose, and dose equivalent—are expressions of the concentration of radiation at some specific point. Two of the quantities—integral exposure and integral dose—tell us something about the total amount of radiation delivered by an x-ray beam.

There are specific units used with each quantity. For example, we use the units of pounds or kilograms for the quantity of weight. We can also express height in either inches or centimeters. Some of our radiation quantities can be expressed in two different units. This is because there is a movement in the world to replace some of the older conventional units with the international system of units (SI units). The adoption of SI radiation units is progressing rather slowly because there is nothing wrong with our conventional units, and SI units are somewhat awkward for a number of common applications. Throughout the text the units used will be those believed to be the most useful to the reader. In this chapter both unit systems are discussed and compared.

Table 4-1 is a listing of the radiation quantities and units most frequently encountered in radiography. It is a useful reference, especially for the conversion of one system of units to another.

We will now consider some of the individual quantities and their associated units.

31

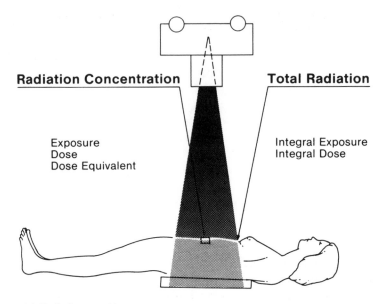

Radiation Concentration   **Total Radiation**

Exposure   Integral Exposure
Dose   Integral Dose
Dose Equivalent

Figure 4-1 Radiation quantities.

EXPOSURE

In radiography an important characteristic of an x-ray beam is the quantity or amount of *exposure* delivered to things like the patient's body, the receptor, and personnel working in the vicinity. Exposure is a quantity that expresses the concentration of radiation delivered to a specific point, rather than the total radiation delivered to the patient's body or other object. Figure 4-1 helps us make this distinction. Imagine that we have drawn a small square on the patient's body at the center of the x-ray beam. Exposure is related to the concentration of radiation photons entering that square. This is different from the total number of photons entering the patient's body throughout the beam area. If we were to change the beam

Table 4-1 Radiation Units and Conversion Factors

Quantity	Conventional Unit	SI Unit	Conversions
Exposure	roentgen (R)	coulomb/kg of air (C/kg)	1 C/kg = 3876 R
			1 R = 258 μC/kg
Dose	rad	gray (Gy)	1 Gy = 100 rad
Dose equivalent	rem	sievert (Sv)	1 Sv = 100 rem

area (FOV), the exposure or concentration in the center of the area would remain the same, but the total radiation entering the body would obviously change.

The Roentgen

The *roentgen* (R) is the conventional unit used in radiography for expressing exposure. In some instances it is more convenient to use the smaller unit, the *milliroentgen* (mR). One roentgen (R) is equal to 1,000 milliroentgens (mR). For example, the exposure to a patient undergoing a radiographic examination could be expressed as either 0.2 R or 200 mR.

Exposure meters are electronic devices used to measure exposure and calibrate radiographic equipment. Most exposure meters use an ionization chamber to measure exposure. The ionization chamber, which is a part of the meter, contains air. When it is exposed to the radiation, the air becomes ionized in proportion to the amount of exposure. Ionization is the process of breaking an atom apart into electrons and ions with a positive charge. Ionization is produced when the x-ray photons interact with atoms. The meter measures the amount of ionization and displays it in units of roentgens or milliroentgens.

There is a new SI unit for expressing radiation exposure, but it is not very popular in radiography. The unit is the coulomb per kilogram of air (C/kg). The relationship between this unit and the roentgen is

$$1 \text{ C/kg} = 3876 \text{ R}$$

Surface Integral Exposure

Since exposure (in roentgens or coulombs per kilogram) is a measure of concentration, it does not express the total amount of radiation delivered to a body. The total radiation delivered, or *surface integral exposure* (SIE), is determined by the exposure and the dimensions of the exposed area.

The surface integral exposure is expressed in the conventional units of *roentgens-square centimeters* (R-cm^2). If the radiation exposure is uniform over the entire area, the SIE is the product of the exposure in roentgens and the exposure area in square centimeters. If the exposure is not the same at all points in the exposed area, the SIE is the sum of the exposure values for each square centimeter of exposed surface. We will see examples later. The SIE can be measured during x-ray examinations by placing a special type of ionization chamber in the path of the x-ray beam. The significance of SIE is that it describes the total radiation imparted to a patient, whereas exposure indicates only the concentration of radiation to a specified point.

Let's now revisit the inverse-square effect that was introduced in Chapter 3 in Figure 3-3. It demonstrates the difference between *exposure* and the *total radiation* (integral exposure) in the beam.

In this example the total radiation (integral exposure) does not change with distance from the x-ray tube. However, it spreads over a larger area as the distance increases, becoming less concentrated. Since exposure is related to the concentration of radiation, exposure decreases as the distance is increased.

The typical fluoroscopy examination provides an excellent opportunity to compare exposure (concentration) and SIE (total radiation). In Figure 4-2 two cases are compared. In both instances the beam area was 10 cm × 10 cm (100 cm^2); the total exposure time was 5 minutes at an exposure rate of 3 R/min. In both instances the SIE is 1,500 R-cm^2. However, the exposure depends on how the x-ray beam is moved during the examination. In the first example the beam was not moved, and the resulting exposure to the skin was 15 R. In the second example the beam was moved to different locations, so that the exposure was distributed over more surface area and the concentration became less.

Another important example is illustrated in Figure 4-3. Here the same exposure (100 mR) is delivered to both patients. However, there is a difference in the exposed area: the patient on the right received 10 times as much radiation as the patient on the left.

The important point to remember is that exposure (roentgens) alone does not express the total radiation delivered to a body. The total exposed area must also be considered.

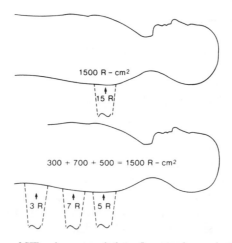

Figure 4-2 Comparison of SIE and exposure during a fluoroscopic examination.

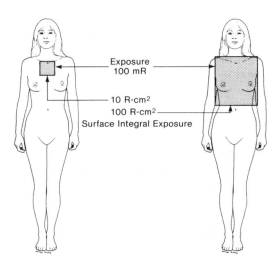

Figure 4-3 Comparison of SIE values for a radiographic examination.

ABSORBED DOSE

A human body absorbs most of the radiation energy delivered to it. The portion of an x-ray beam that is absorbed depends on the penetrating ability of the radiation and the size and density of the body section exposed. In most clinical situations more than 90% is absorbed. Two aspects of the absorbed radiation energy must be considered: the amount (concentration) absorbed at various locations throughout the body and the total amount absorbed.

Absorbed dose is the quantity that expresses the concentration of radiation energy absorbed at a specific point within the body tissue. Since an x-ray beam is weakened by absorption as it passes through the body, all tissues within the path of the beam will not absorb the same dose. The absorbed dose will be much greater for the tissues near the surface where the beam enters than for those deeper within the body. Absorbed dose is defined as the quantity of radiation energy absorbed per unit mass of tissue.

Units

The conventional unit for absorbed dose is the *rad*, which is equivalent to 100 *ergs* of absorbed energy per gram of tissue. The corresponding SI units for the rad and the erg are the gray (Gy) and the joule (J). The *gray* is equivalent to the absorption of 1 *joule* of radiation energy per kg of tissue. The erg and the joule are both units of energy. One joule is equal to 10 million ergs. The relationship between the four units is

$$1 \text{ rad} = 100 \text{ erg/g} = 0.01 \text{ J/kg} = 0.01 \text{ Gy}$$
$$1 \text{ Gy} = 100 \text{ rad}$$

For a specific type of tissue and photon energy spectrum, the absorbed dose is proportional to the exposure delivered to the tissue. However, the ratio of absorbed dose (rads) to exposure (roentgens) is not the same for all types of tissues. The absorbed dose in soft tissue is slightly less than 1 rad per roentgen of exposure throughout the photon energy range used in radiography. In contrast, a bone exposed to 1 R will produce an absorbed dose of approximately 3 rad.

Integral Dose

Integral dose is the total amount of energy absorbed in the body. It depends not only on the absorbed dose values but also the total mass of tissue exposed.

The conventional unit for integral dose is the *gram-rad*, which is equivalent to 100 ergs of absorbed energy. The concept behind the use of this unit is that if we add the absorbed doses (rads) for each gram of tissue in the body, we will have an indication of total absorbed energy.

Integral dose (total absorbed radiation energy) is probably the radiation quantity that most closely correlates with potential radiation damage during a diagnostic procedure. This is because it reflects not only the concentration of the radiation absorbed in the tissue but also the amount of tissue affected by the radiation. There is no practical method for measuring integral dose in the human body.

Computed tomography can be used to demonstrate integral dose, as illustrated in Figure 4-4. We begin with a one-slice examination and assume that the average dose to the tissue in the slice is 5 rad. If there are 400 g of tissue in the slice, the integral dose will be 2,000 gram-rad. If we now perform an examination of 10 slices, but all other factors remain the same, the dose (energy concentration) in each slice will remain the same. However, the integral dose (total energy) increases in proportion to the number of slices and is now 20,000 gram-rad. In this example we made the simplifying assumption that no radiation is scattered from one slice to another. In reality, some radiation is exchanged between neighboring slices, but that does not affect the concept presented.

The same principle applies to other x-ray examinations. Let's go back to the fluoroscopic example in Figure 4-2. The total radiation energy or integral dose is the same for the two cases. However, when the beam is moved so that the radiation

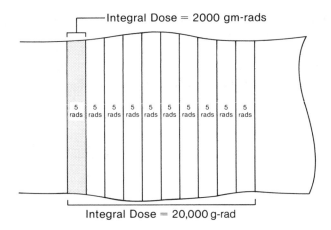

Figure 4-4 Integral dose in computed tomography.

is distributed over more tissue, the maximum dose to any tissue decreases from 15 rads to 7 rads.

Let's now consider the integral dose in Figure 4-3. When the exposed area is increased without changing the exposure value (100 mR), the integral dose is increased because we are exposing more tissue. This is a very important point to remember in radiography. A large FOV deposits more radiation energy (integral dose) in the patient's body than a small FOV, even when the exposure is the same.

BIOLOGICAL IMPACT

It is sometimes desirable to express the actual or relative *biological impact* of radiation. Biological impact is the amount of effect or change in the tissue produced by the radiation. It is necessary to distinguish between the biological impact and the physical quantity of radiation (dose), because all types of radiation do not have the same potential for producing biological change. For example, one rad of one type of radiation might produce significantly more radiation damage than one rad of another type. In other words, biological impact is determined by both the quantity of radiation (dose) and its ability to produce biological effects.

Dose Equivalent

Dose equivalent is the quantity commonly used to express the biological impact of radiation on persons receiving occupational or environmental exposures. Personnel exposure in a clinical facility is often determined (from film badge readings) and recorded as dose equivalent.

Dose equivalent is proportional to the absorbed dose, but also depends on the type of radiation. For x-radiation, the dose equivalent is numerically equal to the absorbed dose. The conventional unit of measure for dose equivalent is the *rem*, and the SI unit is the *sievert* (Sv). An absorbed dose of 1 rad produces a dose equivalent of 1 rem. An absorbed dose of 1 gray produces a dose equivalent of 1 sievert.

Dose equivalent values can be converted from one system of units to the other by

$$1 \text{ Sv} = 100 \text{ rem}$$

Figure 4-5 is a summary of the general relationship among the three quantities: exposure, absorbed dose, and dose equivalent. Although each expresses a different aspect of radiation, they all express radiation concentration. For the types of radiation used in diagnostic procedures, the factors that relate the three quantities have values of approximately 1 in soft tissue. Therefore, an exposure of 1 R produces an absorbed dose of approximately 1 rad, which in turn produces a dose equivalent of 1 rem.

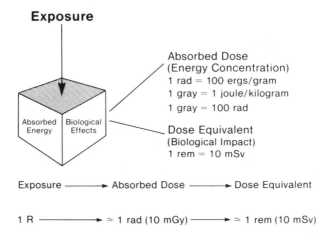

Figure 4-5 Relationship of exposure, absorbed dose, and dose equivalent.

Study Activities

Distinguish between physical quantities and units by giving some examples of each.

Name the conventional unit for expressing exposure.

Name the SI unit for exposure and determine whether it represents a smaller or larger amount of exposure than the roentgen.

Assume that a 10 cm × 10 cm area on a patient's body is exposed to 200 mR. Calculate the SIE.

If the exposure is repeated a second time in the same location, calculate: 1) the total exposure and 2) the total SIE.

If the exposure was repeated but moved to a different area of the patient's body, calculate: 1) the maximum exposure and 2) the total SIE.

Assume that the FOV is 10 cm × 10 cm when the patient is 1 m from the focal spot and the entrance exposure to the patient is 300 mR. If the patient is moved to a distance of 2 m (without changing any other factors), determine the change in: 1) entrance exposure and 2) SIE.

Name the two units that can be used to express absorbed dose.

Determine which absorbed dose unit represents the larger amount of radiation.

Assume that some soft tissue has received an exposure of 2 roentgens. Determine the absorbed dose in both rads and grays.

Explain the difference between absorbed dose and integral dose.

Name the two units for expressing dose equivalent. Indicate which unit represents the larger amount of radiation.

Assume that some tissue has received an exposure of 3 roentgens. Determine the dose equivalent in both units.

Chapter 5

Electrical Energy

An x-ray machine produces x-radiation from electrical energy. We can control both the quantity and quality of radiation by adjusting certain characteristics of the electrical energy that is supplied to the x-ray machine.

Electrical energy is transported from one location to another by very small particles, *electrons*. Electrons move through metal conductors, which are usually in the form of wires. When we turn an x-ray machine on and push the exposure button, many of these small electrons race through the x-ray machine circuits, bringing energy to the x-ray tube. In the x-ray tube, the energy is taken from the electrons and used to create the x-ray photons.

In this chapter we will get to know the electrons better and find out how they carry energy. We will also discover that there are several different quantities and units we must use when setting up an exposure technique.

ELECTRONS AND ENERGY

Electrons are the smallest particles found in matter. An electron has a mass of 9.1×10^{-28}g, which means it would take 10.9×10^{26} electrons to equal the weight of 1 cm^3 of water. The question might be raised as to how such a small particle can be the foundation of our modern technology. The answer is simple— numbers. Tremendous numbers of electrons are involved in most applications. For example, when a 100-watt light bulb is turned on, electrons race through the wires carrying energy to the bulb at the rate of 5.2×10^{18} electrons per second. In addition to its mass, each electron carries a 1-unit negative electrical charge. It is the charge of an electron that enables it to interact with other electrons and particles within atoms.

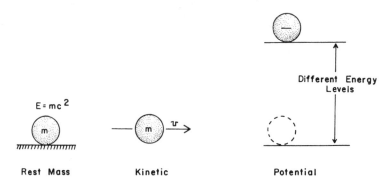

Figure 5-1 Types of energy associated with electrons.

Because an electron has both mass and electrical charge, it can possess energy of several types, as shown in Figure 5-1. It is the ability of an electron to take up, transport, and give up energy that makes it useful in the x-ray system.

The energy of individual electrons, especially in x-ray circuits, is expressed in units of keV. We notice that it is the same unit we use for the energy of individual x-ray photons.

Rest Mass Energy

Even when an electron is at rest and has no apparent motion, it still has energy. In fact, according to the laws of physics, an object has some energy just because of its mass. Under certain conditions, mass can be converted into energy and vice versa. Einstein's famous equation

$$E = mc^2$$

predicts the amount of energy that could be obtained if an object with a mass, m, were completely converted. In this relationship, c is the speed of light. Although it is not possible with our present technology to convert most objects into energy, certain radioactive materials emit particles, called positrons, that can destroy or annihilate electrons. When this happens, the electron's entire mass is converted into energy. According to Einstein's equation, each electron will yield 510 keV. This energy appears as a photon. The annihilation of positrons and electrons is the basis for positron emission tomography (PET). In x-ray imaging we are not concerned with the rest-mass energy of electrons. We have discussed it here just to make the story of electron energy complete. There are two other types of electron energy that play a very important role in the production of x-radiation.

Kinetic Energy

Kinetic energy is associated with motion. It is the type of energy that a moving automobile or baseball has. When electrons are moving they also have kinetic energy related to their velocity (v).

Potential Energy

Potential energy is the type of energy possessed by an object because of its location or configuration and is essentially a relative quantity. That is, an object will have more or less energy in one location or configuration than in another. Although there is generally not a position of absolutely zero potential energy, certain locations are often designated as the zero-energy level for purposes of reference.

Electrons can have two forms of potential energy. One form is related to location within an electrical circuit, and the other is related to location within an atom. Two important aspects of electron potential energy are that energy from some source is required to raise an electron to a higher energy level and that an electron gives up energy when it moves to a lower potential energy position.

Energy Exchange

Because electrons are too small to see, it is sometimes difficult to visualize what is meant by the various types of electron energy. Consider the stone shown in Figure 5-2; we will use it to demonstrate the various types of energy that also apply to electrons.

Potential energy is generally a relative quantity. In this picture the ground level is arbitrarily designated as the zero potential energy position. When the stone is raised above the ground, it is at a higher energy level ($+E$). If the stone is placed in a hole below the surface, its potential energy is negative with respect to the ground level. However, its energy is still positive with respect to a position in the bottom of a deeper hole. The stone at position A has zero potential energy (relatively speaking) and zero kinetic energy because it is not moving. When the man picks up the stone and raises it to position B, he increases its potential energy with respect to position A. The energy gained by the stone comes from the man. We show later that electrons can be raised to higher potential energy levels by devices called power supplies or generators. The additional potential energy possessed by the stone at B can be used for work or can be converted into other forms of energy. If the stone were connected to a simple pulley arrangement and allowed to fall back

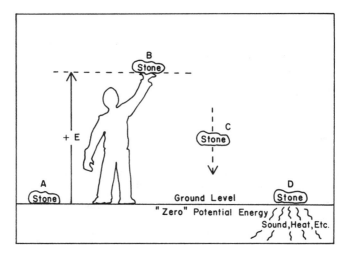

Figure 5-2 Transfer of energy from one form to another.

to the ground, it could perform work by raising an object fastened to the other end of the rope.

If the man releases the stone at B and allows it to fall back to the ground, its potential energy is converted into kinetic energy. As the stone moves downward, decreasing its potential energy (which is proportional to its distance above the ground), it constantly increases its speed and kinetic energy. Just before it hits the ground, its newly gained kinetic energy will be just equal to the potential energy supplied by the man. Electrons undergo a similar process within x-ray tubes where they swap potential for kinetic energy. Just as the stone reaches the surface of the ground, it will have more total energy than it had when it was resting at position A. However, when it comes to rest on the ground at D, its potential energy level is the same as at A. The extra energy originally supplied by the man must be accounted for. In this situation, this energy is converted into other forms, such as sound, a small amount of heat, and mechanical energy used to alter the shape of the ground. When high-speed electrons strike certain materials, they also lose their kinetic energy; their energy is converted into heat and x-radiation.

Energy Transfer

One of the major functions of electrons is to transport energy from one location to another. We have just seen that individual electrons can possess several forms of energy. The principle of electrical energy transportation is that electrons pick up energy in one location and then move to another, where they pass the energy on to some other material. Generally the arrangement is such that the electrons then move back to the energy source and repeat the process.

The pathway electrons travel as they transfer energy from one point to another is a *circuit*. A basic electrical circuit is shown in Figure 5-3. All circuits must contain at least two components (or devices), as shown. One component, designated here as the source, can convert some other type of energy (chemical, mechanical, etc) and transfer it to the electrons. Batteries are good examples of electron energy sources. Batteries convert chemical energy into electrical energy. Electrical energy sources such as batteries do not actually create new electrons. They take in electrons that have a very low energy at one terminal and then add additional energy to them before they are expelled from the other terminal. The other component, designated here as a load, performs essentially the opposite function. As the electrons pass through the device, they lose their energy as it is converted into some other form. A light bulb is a good example of a load in which electron energy is converted into light and heat. An x-ray tube is also a load device that converts electrical energy into two other forms: x-radiation and heat.

The energy source and the load are connected with two *conductors* over which the electrons can freely move. The ideal conductor offers no resistance to the flow of the electrons. If the conductor offers significant resistance, the electrons lose some of their energy there. The lost energy is converted into heat.

Electrical circuits neither create nor destroy electrons. The electrons are always present within the conductive materials. Energy is given to and taken from the electrons as they move around the circuit.

The energy carried by the electrons is a form of potential energy. Even though the electrons are moving through the conductors, their velocity is not sufficient to

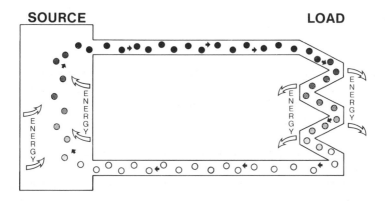

Figure 5-3 A basic electrical circuit.

give them significant kinetic energy. When electrons are moving through free space, like in an x-ray tube, they can carry significant kinetic energy, but they cannot when they are moving through solid conductors. In the typical electrical circuit, one conductor has higher potential energy than the other conductor. In principle, the energy source elevates the electrons to a higher potential energy level, which they maintain until they give up the energy in passing through the load device. The electrons at the lower potential level return to the energy source to repeat the process.

The connection points (terminals) between the source and load devices and the conductors have a polarity and are designated as either positive or negative. Polarity means opposite characteristics. The electrons exit the source at the negative terminal and enter the negative terminal of the load. They then exit the positive terminal of the load device and enter the source at the positive terminal. In principle, the negative conductor contains the electrons at the high potential energy level. The positive conductor contains the electrons that have lost their energy and are returning to the source. In *direct current* (DC) circuits the polarities do not change. However, in *alternating current* (AC) circuits the polarity of the conductors is constantly alternating between negative and positive.

ELECTRICAL QUANTITIES

Each electron passing through the circuit carries a very small amount of energy. However, by collective effort electrons can transport a tremendous amount of energy. The amount of energy transferred by an electrical circuit depends on the quantity of electrons and the energy carried by each. We now consider these specific electrical quantities and their associated units.

Current

When an electrical circuit is in operation, electrons are continuously moving or flowing through the conductor. The flowing electrons are known as an *electrical current*. The number of electrons that move past a given point per second is a measure of the current. Since, in the typical circuit, the number of electrons per second is quite large, a more useful unit than this number is desirable. The basic unit of current is the *ampere* (A). One ampere is defined as the flow of 6.25×10^{18} electrons per second. In x-ray machines the current is typically a fraction of 1 A, so that the *milliampere* (mA) is a more appropriate unit. As indicated in Figure 5-4, a current of 1 mA is equal to the flow of 6.25×10^{15} electrons per second past a given point. The current that flows through an x-ray tube is generally referred to as the *MA*, referring to the quantity of current. The unit of measure, milliampere, is written as mA.

CURRENT

1 mA = 6.25 × 10^{15} electrons per second

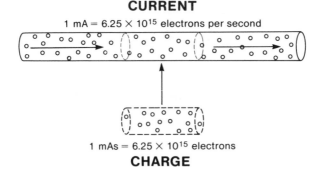

1 mAs = 6.25 × 10^{15} electrons

CHARGE

Figure 5-4 Electrical current and charge.

Electron Quantity and Charge

In addition to the rate at which electrons are flowing through a circuit (ie, the current), it is often necessary to know the total quantity of electrons in a given period of time. In x-ray work the most appropriate measurement unit for specifying electron quantity is the milliampere-second (mAs). The total quantity of electrons passing through the x-ray tube, the MAS, is the product of the current (MA) and the time in seconds (S). Since a current of 1 mA is a flow of 6.25 × 10^{15} electrons per second, it follows that 1 mAs is a cluster of 6.25 × 10^{15} electrons, as shown in Figure 5-4.

Voltage

We pointed out earlier that electrons could exist at different potential energy levels because of either their different positions within the atom or their different locations within an electrical circuit. Consider the two wires or conductors shown in Figure 5-5. The electrons in one of the conductors are at a higher potential energy level than the electrons in the other. Generally, the electrons in the negative conductor are considered to be at the higher energy level. An electrical quantity that indicates the difference in electron potential energy between two points within a circuit is the *voltage*, or potential difference. The unit used for voltage, or potential difference, is the *volt* (V). The difference in electron potential energy between two conductors is directly proportional to the voltage. Each electron will have an energy difference of 1 electron volt (eV) for each volt. An electron volt is the quantity of energy that an electron gains or loses, depending on direction, when it moves between two points in a circuit that have a 1 V difference. In the basic x-ray machine circuit, the voltage is on the order of thousands of volts

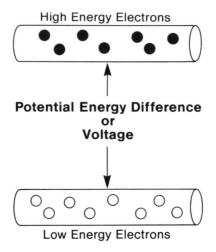

High Energy Electrons

Potential Energy Difference
or
Voltage

Low Energy Electrons

Figure 5-5 Electron potential energy or voltage.

(kilovolts) and is often referred to as the KV or KV_p, meaning the quantity of voltage or potential. The matching unit of measurement is written as kV or kV_p.

Power

Power is the quantity that describes the rate at which energy is transferred. The *watt* is the unit of power and is equivalent to an energy transfer rate of 1 J/sec. A joule is a relatively large unit of energy. The power in an electrical circuit is proportional to the energy carried by each electron (voltage) and the rate of electron flow (current). The specific relationship is

$$\text{Power (watts)} = \text{Voltage (volts)} \times \text{Current (amperes)}$$

Total Energy

The amount of energy that an electrical circuit transfers depends on the voltage, the current, and the duration of the energy transfer. The fundamental unit of energy is the joule. The relationship of total transferred energy to the other electrical quantities is

$$\text{Energy (joules)} = \text{Voltage (volts)} \times \text{Current (amperes)} \times \text{Time (seconds)}$$

In a later chapter we will make use of our knowledge of power and energy to determine the amount of undesirable heat produced in an x-ray tube.

THE X-RAY CIRCUIT

The basic circuit shown in Figure 5-6 is found in all x-ray machines. The power supply (generator) that gives energy to the electrons and pumps them through the circuit is discussed in detail in Chapter 7. The voltage between the two conductors in the x-ray circuit is typically in the range of 30,000 to 120,000 V (30 to 120 kV). Voltage is generally adjustable, and an appropriate value can be selected by the operator of the x-ray equipment.

In this circuit the x-ray tube is the load. It is the place where the electrons lose their energy. The energy lost by electrons in passing through an x-ray tube is converted into heat and x-ray energy.

Alternating Current

In some electrical circuits, the voltage and current remain constant with respect to time, and the current always flows in the same direction. These are generally designated as direct current (DC) circuits. A battery is an example of a power supply that produces a direct current.

Some power supplies, however, produce voltages that constantly change with time. Since in most circuits the current is more or less proportional to the voltage, it also changes value. In most circuits of this type, the voltage periodically changes polarity, and the current changes or alternates direction of flow. This is an alternating current (AC) circuit. The electricity distributed by power companies is AC. There are certain advantages to AC, in that devices can be used for increasing or decreasing voltages, and also many motors are designed for AC operation.

If a graph of the changing values of either the AC voltage or the current is plotted with respect to time, it will generally be similar to the one shown in Figure 5-7.

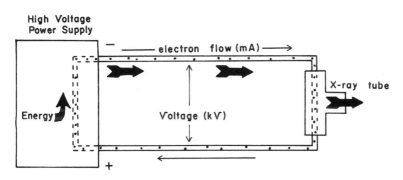

Figure 5-6 The x-ray circuit.

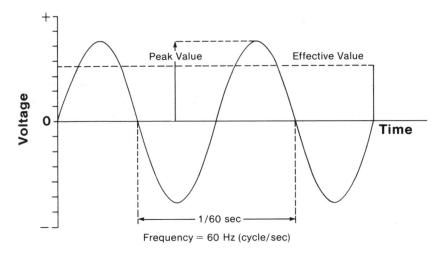

Figure 5-7 Waveform of an alternating voltage.

This representation of the voltage with respect to time is known as the *waveform*. Most AC power sources produce voltages with the sine waveform, shown in Figure 5-7. This name is derived from the mathematical description of its shape.

One characteristic of an alternating voltage is its *frequency*. The frequency is the rate at which the voltage changes through one complete cycle. The time of one complete cycle is the period; the frequency is the reciprocal of the period. For example, the electricity distributed in the United States goes through one complete cycle in 0.0166 seconds and has a frequency of 60 cycles per second. The unit for frequency is the hertz (Hz), which is 1 cycle per second.

During one voltage cycle, the voltage changes continuously. At two times during the period it reaches a peak but remains there for a very short time. This means that for most of the period the circuit voltage is at less than the peak value. For the purpose of energy and power calculations, an effective voltage value, rather than the peak value, should be used. For the sine-wave voltage, the effective value is 70.7% (0.707) of the peak voltage. This is the waveform factor, and its value depends on the shape of the voltage waveform.

Study Activities

Name the unit used to express the energy of individual electrons.

Describe the condition that gives a baseball kinetic energy.

Describe the condition that gives a baseball potential energy.

Name the unit of measure for current.

Determine the number of milliamperes for a current of 0.5 ampere.

Calculate the time required to produce 50 mAs if the current is 300 mA.

Calculate the number of electrons passing through a circuit in 2 seconds if the current is 10 mA.

Name the unit of measure for expressing the difference in electrical potential energy.

Calculate the power of a lightbulb operating with 120 volts and a current of 0.83 ampere.

Calculate the current through a 50-watt lightbulb operating at 120 volts.

Explain the difference between voltage and current.

Explain the difference between alternating current (AC) and direct current (DC).

Explain the difference between the peak and effective values of an alternating voltage.

Chapter 6

Physical Characteristics of Matter

The physical universe consists of two things: energy and matter. Energy can be in many different forms. We have already learned about some of the forms of energy that we encounter in radiography. Examples are x-radiation, light, heat, and electrical energy. We will now turn our attention to the physical characteristics of matter that are important in the practice of radiography.

Matter is anything that has mass or weight. The earth consists of matter, and so do all of the objects on it. That includes our own bodies and everything that we can touch. Every part of a radiographic system consists of some form of matter. Matter plays an important role in both producing and absorbing the x-ray beam. X-radiation is produced within the matter that makes up the anode of the x-ray tube. It is absorbed in the matter of such objects as the patient's body, intensifying screens, and lead shields.

FORMS OF MATTER

Matter comes in many different forms. All matter sharing the same general properties is known as a *substance*. Examples of substances are water, iron, salt, air, and iodine. Substances can be in the form of pure chemical elements (iron and iodine) or several elements chemically combined to form compounds (water, salt, sugar).

Much of the matter in the world consists of a *mixture* of elements or compounds, which are not all chemically combined. Examples of mixtures are tissue, air, wood, and soil.

The name of a substance is the name of the element, compound, or mixture that makes up the substance.

ELEMENTS AND ATOMS

The simplest form of a substance is a chemical *element*. The elements are the basic building blocks of the other two forms of matter, compounds and mixtures. There are only 108 different chemical elements known at the present time. One hundred of these elements are listed in Table 6-1. The ones not listed are very rare and do not play a significant role in our lives. If all substances were broken down to the basic elements, then we would only have approximately 100 different substances.

What makes each element different from another is the structure of its atoms. All atoms of any one element—for example, carbon—are similar. However, all carbon atoms are different from those of the other chemical elements.

Atomic Structure

An atom is the smallest unit of substance that can still be identified as a specific chemical element. For example, if we should take one carbon atom and break it apart, it would no longer be carbon.

The basic structure of an atom is illustrated in Figure 6-1. All atoms are built with three components: *protons, neutrons,* and *electrons.* The protons and neutrons are clustered together and form the *nucleus* of the atom. They are much larger than the electrons. This means that most of the mass of an atom is concentrated in the nucleus. The protons have positive electrical charges, the electrons have negative charges, and the neutrons are neutral, with no charge. The electrons are located in *shells*, orbits that surround the nucleus.

Atomic Number

Each chemical element has its unique atomic number (Z). This is the number of protons in its nucleus. The smallest possible atom is the one with a nucleus consisting of a single proton. Such a nucleus has an atomic number of 1, which identifies it as hydrogen. Look at Table 6-1 and notice that the elements are arranged in order of the atomic number. The atomic number indicates the approximate size of the individual atoms. Hydrogen atoms are very small. Gold, which has an atomic number of 79, is a very large atom because its nucleus contains 79 protons compared to the 1 proton in hydrogen.

The number of neutrons in a nucleus is approximately the same as the number of protons. There is some variation, which produces the different isotopes (forms) of an element, but that is not of any concern in radiography.

Table 6-1 The First 100 Chemical Elements and Their Atomic Numbers

Atomic Number	Element	Symbol	Atomic Number	Element	Symbol	Atomic Number	Element	Symbol	Atomic Number	Element	Symbol
1	Hydrogen	H	26	Iron	Fe	51	Antimony	Sb	76	Osmium	Os
2	Helium	He	27	Cobalt	Co	52	Tellurium	Te	77	Iridium	Ir
3	Lithium	Li	28	Nickel	Ni	53	Iodine	I	78	Platinum	Pt
4	Beryllium	Be	29	Copper	Cu	54	Xenon	Xe	79	Gold	Au
5	Boron	B	30	Zinc	Zn	55	Cesium	Cs	80	Mercury	Hg
6	Carbon	C	31	Gallium	Ga	56	Barium	Ba	81	Thallium	Tl
7	Nitrogen	N	32	Germanium	Ge	57	Lanthanum	La	82	Lead	Pb
8	Oxygen	O	33	Arsenic	As	58	Cerium	Ce	83	Bismuth	Bi
9	Fluorine	F	34	Selenium	Se	59	Praseodymium	Pr	84	Polonium	Po
10	Neon	Ne	35	Bromine	Br	60	Neodymium	Nd	85	Astatine	At
11	Sodium	Na	36	Krypton	Kr	61	Promethium	Pm	86	Radon	Rn
12	Magnesium	Mg	37	Rubidium	Rb	62	Samarium	Sm	87	Francium	Fr
13	Aluminum	Al	38	Strontium	Sr	63	Europium	Eu	88	Radium	Ra
14	Silicon	Si	39	Yttrium	Y	64	Gadolinium	Gd	89	Actinium	Ac
15	Phosphorus	P	40	Zirconium	Zr	65	Terbium	Tb	90	Thorium	Th
16	Sulfur	S	41	Niobium	Nb	66	Dysprosium	Dy	91	Protoactinium	Pa
17	Chlorine	Cl	42	Molybdenum	Mo	67	Holmium	Ho	92	Uranium	U
18	Argon	Ar	43	Technetium	Tc	68	Erbium	Er	93	Neptunium	Np
19	Potassium	K	44	Ruthenium	Ru	69	Thulium	Tm	94	Plutonium	Pu
20	Calcium	Ca	45	Rhodium	Rh	70	Ytterbium	Yb	95	Americium	Am
21	Scandium	Sc	46	Palladium	Pd	71	Lutetium	Lu	96	Curium	Cm
22	Titanium	Ti	47	Silver	Ag	72	Hafnium	Hf	97	Berkelium	Bk
23	Vanadium	V	48	Cadmium	Cd	73	Tantalum	Ta	98	Californium	Cf
24	Chromium	Cr	49	Indium	In	74	Tungsten	W	99	Einsteinium	Es
25	Manganese	Mn	50	Tin	Sn	75	Rhenium	Re	100	Fermium	Fm

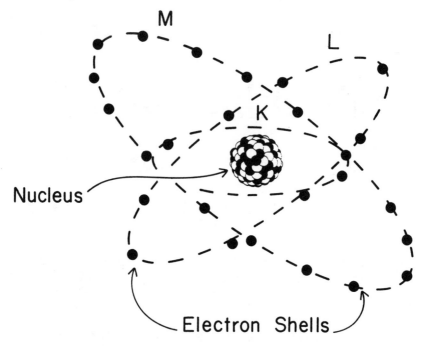

Figure 6-1 Structure of an atom.

The number of electrons in a normal atom is the same as the number of protons in its nucleus. For example, a carbon atom, which always has an atomic number of 6 (six protons), will normally have six electrons surrounding the nucleus.

Ions

When an atom has the same number of negatively charged electrons as positively charged protons, the charges produce a balance so that the overall electrical charge of the atom is zero. This is the normal condition. However, if an atom should either lose or gain an electron, the charge balance is upset, and the atom will then have an electrical charge. An atom in this condition is known as an *ion*. If an atom loses one of its electrons, it becomes a positive ion, because it has one positive proton that is no longer balanced by a negative electron. If an atom gains an extra electron, it becomes a negative ion.

When x-ray photons interact with matter, they usually do so by knocking some of the electrons out of the atoms. This is why x-radiation is classified as *ionizing radiation*.

We have already seen that the electrons are located in shells (orbits) surrounding the nucleus. Most atoms, if they are large enough, have several different electron shells. The different shells are located different distances from the nucleus. The shell closest to the nucleus is named the K-shell. The next is the L-shell. If the atom is large enough, it will have additional shells designated as the M-shell, the N-shell, etc., in alphabetical order.

Each shell has a limited capacity for electrons. The K-shell cannot hold more than two electrons. The L-shell has a capacity for a maximum of eight electrons. This means that the electrons in a typical atom are not bunched together in one shell but are distributed among several shells. Atoms are built up by filling the innermost K-shell and then working outward.

Binding Energy

The electrons are bound to the atom by their attraction to the positive nucleus. Remember that unlike electrical charges (positive, negative) attract each other. One important characteristic of an electron is how strongly it is bound to the atom. The strength of an electron's binding to an atom is its *binding energy*. Binding energy is not energy that the electron actually has. It is the amount of energy that would be required to pull the electron out of the atom. An electron with a large binding energy will be much harder to pull or knock out of an atom than an electron with a small binding energy.

The amount of binding energy for an individual electron depends on two things. One is the shell in which it is located. Electrons in the K-shell, which are the closest to the nucleus, have a much higher binding energy than electrons in the more distant shells. The other factor is the size (atomic number) of the atom. The binding is much stronger in large atoms than small atoms.

In radiography we will be concerned with the binding energy of the K-shell electrons of several different elements. This is because the binding energy has a very large effect on the ability of different substances to produce and absorb x-radiation. It is not necessary for us to know precise binding energy values. However, there will be several occasions when we will apply the idea of binding energy. The K-shell binding energy for several important elements is shown in Figure 6-2.

COMPOUNDS AND MIXTURES

Many substances are composed of more than one chemical element. There are two ways in which elements are combined to form other substances. One is a chemical process and the other a physical process.

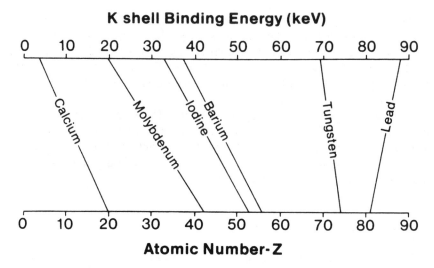

Figure 6-2 Relationship between K-shell binding energy and atomic number.

When two or more atoms are chemically bonded together to form a molecule, the substance is known as a compound. Water is a good example of a compound. A water molecule contains two hydrogen atoms attached to one atom of oxygen.

When two or more elements are together but the atoms are not chemically joined together to form molecules, the substance is a mixture. Air is a good example of a mixture. Air is a mixture of several different elements and compounds. The most prevalent substances in air are the elements nitrogen, oxygen, and argon and the compound carbon dioxide. A mixture does not have a distinct chemical characteristic of its own. Each element or compound in a mixture is free to engage in a chemical interaction. For example, when there is a fire, the oxygen in air combines with the burning substance to create more carbon dioxide. However, the other elements, such as nitrogen and argon, don't get involved in the process.

A compound has distinct chemical characteristics of its own, which are often quite different from the chemical characteristics of the individual elements making up the compound.

Only the chemical elements have exact atomic numbers. Since both compounds and mixtures contain a variety of elements, they do not have their unique atomic numbers. However, a compound or a mixture will have an effective atomic number, which is determined by combining the atomic numbers of the elements that it contains. Later we will see some values for the effective atomic number of several different substances encountered in radiography.

DENSITY AND SPECIFIC GRAVITY

Density is one of the most important physical properties of substances involved in the process of radiography. The density of a substance is the mass of a 1-unit volume of the substance. The unit volume most frequently used for specifying density is the cubic centimeter (cc or cm^3). A cubic centimeter is like a little square box with each edge one centimeter in length. A volume of 1 cc is the same as 1 milliliter (ml). The conventional unit for mass is the gram. Therefore, the density of a substance is the number of grams per cc (g/cc). Pure water has a density of 1 g/cc. Substances that are less dense than water, such as oil, will float. Substances that are more dense, such as lead, will sink.

Most substances have reasonably specific density values. However, the density of some substances, especially gases, might depend on temperature and pressure. The density of a liquid that is a mixture of water and some other substance depends on the concentration. In fact, this is often used as a method for measuring the concentration of substances in a solution.

Specific Gravity

The density of liquids is often measured and described in terms of *specific gravity*. Specific gravity is the same as density except for the units of measurement after the number. Whereas density is expressed in units of g/cc, specific gravity has no units. Water, which has a density of 1 g/cc, has a specific gravity of just 1. Specific gravity is the same as density; it just does not have a unit of measurement.

The specific gravity of a solution is measured by floating a *hydrometer* in it. The hydrometer has a scale that measures how well the instrument floats in the solution. This is determined by the specific gravity of the solution. More of the hydrometer will float above the surface in a solution with a high density (specific gravity) than in a solution with low density. The specific gravity of the solution is read directly from the scale on the hydrometer.

In an x-ray facility the specific gravity of the film developer solution can be measured to determine if it has been mixed to the correct concentration.

SUMMARY

Many different substances are involved in the process of creating and absorbing x-radiation. The role of a specific substance depends on its physical characteristics. Three of the most important characteristics are atomic number, electron binding energy, and density. These characteristics are summarized in Table 6-2 for many of the substances involved in radiography.

Table 6-2 Physical Characteristics of Materials Involved in Photon Interactions

Material	Atomic number* (Z)	K electron binding energy (keV)	Density (g/cc)	Application
Beryllium	4.0		1.85	Low absorbing tube window
Fat	5.92		0.91	Body tissue
Water	7.42		1.0	Tissue "equivalent"
Muscle	7.46		1.0	Body tissue
Air	7.64		0.00129	
Aluminum	13.0		2.7	X-ray filter and penetration reference
Bone (femur)	14.0		1.87	Body tissue
Calcium	20.0		1.55	Body deposits
Copper	29.0	8.9	8.94	X-ray filter
Molybdenum	42.0	20.0	10.22	X-ray source
Silver	47.0	25.5	10.5	Absorber in film
Iodine	53.0	33.2	4.94	Contrast medium and receptor absorber
Xenon	54.0	34.5	0.0059	Receptor absorber
Barium	56.0	37.4	3.5	Contrast medium and receptor absorber
Lanthanun	57.0	38.9	6.15	Receptor absorber
Gadolinium	64.0	50.2	7.95	Receptor absorber
Tungsten	74.0	69.5	19.3	X-ray source
Lead	82.0	88.0	11.34	X-ray absorber for shielding

*Effective Z of tissues from Spiers (1946).

Source: Spiers FW. Effective atomic number and energy absorption in tissues. *Br J Radiol* 1946;19:218.

Study Activities

Draw a diagram of a carbon atom and label the parts.

Name five common chemical elements.

Name five common chemical compounds.

Name five common substances that are mixtures.

Describe an ion.

Determine the density of a substance if a 10-cc volume has a mass of 7 grams.

Explain whether and why the previous substance will sink or float in water.

Name the electron shell that contains the electrons with the greatest binding energy.

Describe the general relationship between electron binding energy and atomic number.

State the number of electrons contained in an atom that has six protons and eight neutrons.

Name the instrument that can be used to measure the specific gravity of a fluid.

Chapter 7

X-Ray Production

X-radiation is created in the x-ray tube, where the electrical energy coming from the generator is converted into x-ray photons. The quantity and quality of the x-radiation can be controlled by adjusting the two electrical quantities, KV and MA, and the exposure time, S. In this chapter, we first become familiar with the design and construction of x-ray tubes, then look at the x-ray production process.

THE X-RAY TUBE

Function

An x-ray tube is an energy converter. It receives electrical energy and converts it into two other forms: x-radiation and heat. The heat is an undesirable by-product. X-ray tubes are designed and constructed so that they can tolerate the heat and dissipate it as rapidly as possible.

Figure 7-1 is a cross-sectional view of a typical tube. This is the glass *insert*, which we usually do not see because it is contained in the metal housing. The space within the insert is a vacuum. The x-ray tube is a relatively simple electrical device, typically containing only two elements: a *cathode* and an *anode*. The anode in most radiographic tubes spins or rotates at a very high speed to spread the heat over a large area. The rotor shown in the illustration is part of an electric motor assembly that spins the anode.

Anode

The anode is the component in which the x-radiation is produced. It is a relatively large piece of metal that has two primary functions: (1) to convert

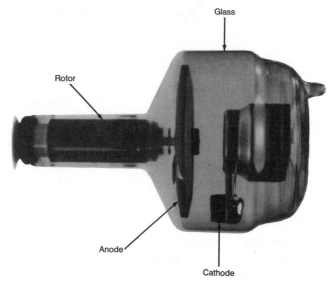

Figure 7-1 Cross-section of a typical x-ray tube.

---

electronic energy into x-radiation and (2) to dissipate the heat created in the process.

Most radiographic procedures use x-ray tubes with tungsten anodes. One exception is mammography, which is often performed with tubes that have molybdenum anodes, for reasons we will discuss later.

Focal Spot

Not all of the anode is involved in x-ray production. The radiation is produced in a very small area on the surface of the anode, the focal spot. X-ray tubes are designed to have specific focal spot sizes, as we have already observed.

The focal spot is the actual area on the surface of the anode that is hit by a beam of electrons coming from the cathode.

Cathode

The basic function of the cathode is to expel the electrons from the electrical circuit and focus them into a well-defined beam aimed at the focal spot on the anode. The typical cathode consists of a small coil of wire (a filament) recessed within a cup-shaped region, as shown in Figure 7-2.

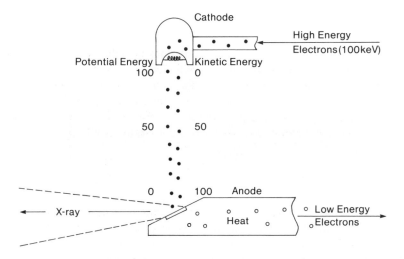

Figure 7-2 Energy exchanges within an x-ray tube.

Electrons that flow through electrical circuits cannot generally escape from the conductor material and move into vacuum space. They can, however, if they are given sufficient energy. One way of doing this is to heat the metal filament to a very high temperature. The thermal energy (or heat) expels the electrons from the cathode in a process known as *thermeonic emission*. The filament of the cathode is heated in the same way as a light bulb filament, by passing an electrical current through it. This heating current is not the same as the current flowing through the x-ray tube that produces the x-radiation.

Envelope

The anode and cathode are contained in an airtight vacuum enclosure, the *glass envelope*. The envelope and its contents are the *tube insert*. This is the part that is replaced when a tube becomes defective. The majority of x-ray tubes have glass envelopes, although tubes for some applications have metal and ceramic envelopes.

The primary functions of the envelope are to provide support and electrical insulation for the anode and cathode assemblies and to maintain a vacuum in the tube. The presence of gases (air) in the x-ray tube would allow electricity to flow through the tube freely, rather than only in the electron beam. This would interfere with x-ray production and possibly damage the circuit. One way in which x-ray tubes become defective is by becoming "gassy."

Housing

The x-ray tube housing serves several functions in addition to enclosing and supporting the other components. It is a shield that absorbs radiation, except for the radiation that passes through the window as the useful x-ray beam. Its relatively large exterior surface expels most of the heat created within the tube. The space between the housing and the insert is filled with oil, which provides electrical insulation and transfers heat from the insert to the housing surface.

ELECTRON ENERGY

The energy that will be converted into x-radiation (and heat) is carried to the x-ray tube by a current of flowing electrons. As the electrons pass through the x-ray tube, they undergo two energy conversions, as illustrated in Figure 7-2. The electrical potential energy is first converted into kinetic energy, which is, in turn, converted into x-radiation and heat.

Potential Energy

When the electrons arrive at the x-ray tube, they carry electrical potential energy. The amount of energy carried by each electron is determined by the voltage (KV) between the anode and cathode. For each kilovolt of voltage, each electron has one kiloelectronvolt of energy. By adjusting the KV, the x-ray machine operator actually assigns a specific amount of energy to each electron.

Kinetic Energy

After the electrons are emitted from the cathode, they come under the influence of an electrical force pulling them toward the anode. This force accelerates them, causing them to move faster and faster and to increase their kinetic energy. This increase in kinetic energy continues as the electrons travel from the cathode to the anode. However, as the electron moves from cathode to anode, its electrical potential energy is decreasing because of its being converted into the kinetic form. Just as the electron arrives at the surface of the anode, its potential energy is lost, and all its energy is kinetic. At this point the electron is traveling with a high velocity, determined by its actual energy content. A 100-keV electron reaches the anode surface traveling at more than one half the velocity of light. When the electrons strike the surface of the anode, they are stopped very quickly and lose their kinetic energy; the kinetic energy is converted into either *x-radiation* or *heat*.

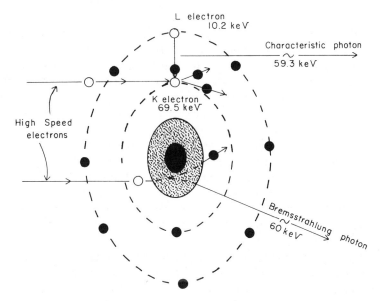

Figure 7-3 Electron-atom interactions that produce x-ray photons.

The electrons interact with individual atoms of the anode material, as shown in Figure 7-3. Two types of interactions produce radiation. An interaction with electron shells produces *characteristic x-ray photons*; interactions with the atomic nucleus produce *Bremsstrahlung x-ray photons*.

BREMSSTRAHLUNG

Production Process

The interaction that produces the most photons is the Bremsstrahlung process. Bremsstrahlung is a German word for "breaking radiation" and is a good description of the process. Electrons that penetrate the anode material and pass close to a nucleus are deflected and slowed down by the attractive force from the nucleus. The energy loss by the electron during this encounter appears in the form of an x-ray photon. All electrons do not produce photons of the same energy.

Spectrum

Only a few of the x-ray photons will have energies close to that of the electrons; most will have lower energies. Although the reason for this is complex, a simplified model of the Bremsstrahlung process is shown in Figure 7-4. First, assume that there is a space, or field, surrounding the nucleus, in which the electrons experience the "breaking" force. This field can be divided into zones, as illustrated. This gives the nuclear field the appearance of a target, with the actual nucleus being located in the center. An electron striking anywhere within the target experiences some breaking action and produces an x-ray photon. Those electrons striking near the center are subjected to the greatest force and, therefore, lose the most energy and produce the highest-energy photons. The electrons hitting in the outer zones experience weaker interactions and produce lower energy photons. Although the zones have essentially the same width, they have different areas. The area covered by a given zone depends on its distance from the nucleus. Since the number of electrons hitting in a given zone depends on its total area, it is obvious that the outer zones capture more electrons and create more photons. From this model an x-ray energy spectrum, such as the one shown in Figure 7-4, could be predicted.

The maximum energy that an x-ray photon can have is equal to the energy carried by each of the electrons. It is not possible for one or more electrons to add their energies together and produce one photon. Only a few photons are produced with maximum energy. This is 70 keV for the example shown. Below this point, the number of photons produced increases as photon energy decreases. The spectrum emerging from the tube generally looks quite different from the one shown here, because of filtration.

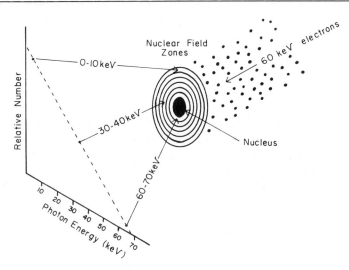

Figure 7-4 A model for Bremsstrahlung production and the associated photon energy spectrum.

A significant number of the low-energy (soft) photons are absorbed or filtered out as they attempt to pass through the anode surface, the x-ray tube window, or added filter material. This filtration effect is good, because these low-energy photons are not capable of penetrating the patient's body and contributing to the image; they only expose the patient to x-radiation.

KV

The high-energy end of the spectrum (maximum photon energy) is determined by the kilovoltage (KV) applied to the x-ray tube. This is because the KV establishes the energy of the electrons as they reach the anode, and no x-ray photon can be created with an energy greater than that of an individual electron. The maximum photon energy, therefore, in kiloelectron volts is numerically equal to the maximum applied potential in kilovolts. In some x-ray equipment the KV might vary during the exposure. The maximum photon energy is determined by the maximum, or peak, voltage during the exposure time. This value is generally referred to as the *kilovolt peak* (KV_p) and is one of the adjustable technique factors.

In addition to establishing the maximum x-ray photon energy and the quality of the beam, the KV has a major role in determining the quantity of radiation (exposure) produced by the tube. High-energy electrons are more efficient in producing radiation by the Bremsstrahlung process than lower-energy electrons.

Changing the KV_p will generally alter the Bremsstrahlung process, as shown in Figure 7-5. The total area under each spectrum curve represents the quantity of radiation (exposure) produced. Notice that as the KV is increased (from 50 to 100 KV), two things happen: (1) The quantity of radiation (exposure) is increased, and (2) the quality of radiation is increased because more photons are produced at the higher energies.

CHARACTERISTIC RADIATION

Production Process

Characteristic radiation is a form of x-radiation produced by a collision between the high-speed electrons from the cathode and the electrons that are part of the anode material atoms. This is illustrated in Figure 7-3. It is a somewhat complicated process, which we will not consider in any great detail. What happens is that the electrons striking the anode material knock out some of the electrons that are part of the anode material atom. When the atom begins to rearrange its electrons to fill in the vacant positions, x-ray photons are produced.

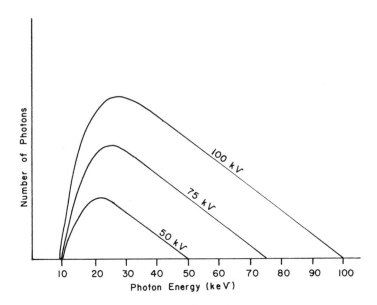

Figure 7-5 Comparison of photon energy spectra produced at different KV_p values.

Characteristic radiation is produced with just a few specific photon energies, unlike Bremsstrahlung radiation, which has many photon energies. The actual photon energies produced depends on the anode material. In fact, that is why it is referred to as characteristic radiation. Tungsten anodes can produce characteristic radiation with photon energies of 57, 59, 67, and 69 keV. Molybdenum anode tubes produce photons with energies of 17.9 and 19.5 keV.

Characteristic radiation is produced in addition to Bremsstrahlung radiation. Figure 7-6 shows a spectrum for an x-ray beam that contains both Bremsstrahlung and characteristic radiation. The characteristic radiation appears as small peaks or lines on top of the Bremsstrahlung spectrum.

KV

The KV value selected by the operator has a strong influence on the production of characteristic radiation. No characteristic radiation will be produced unless the KV_p is set to a value somewhat greater than the energies of the characteristic x-ray photons. For example, with tungsten anode tubes no significant characteristic radiation is produced at settings less than approximately 70 kV_p. As the KV is increased above this value, the quantity (but not the photon energies) of the characteristic radiation is increased.

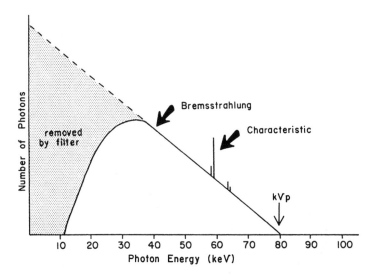

Figure 7-6 Typical photon energy spectrum from a machine operating at $KV_p = 80$.

EFFICIENCY

Only a small fraction of the electrical energy delivered to the anode is converted into x-radiation; most is absorbed and converted into heat. The *efficiency* of x-ray production is the fraction of the total energy that converted into x-radiation. For most radiographic procedures this is in the order of 1% (most of the energy goes into heat). A high efficiency is desirable because it means we can get sufficient radiation to produce an image without being limited by x-ray tube heating.

There are two factors that have an effect on x-ray production efficiency. One of these is the anode material; since most anodes are tungsten, we have very little control over this. The second factor, and one that we can control, is the KV. X-ray production efficiency increases with KV. In other words, a high KV technique produces more radiation in proportion to the heat than a low KV technique.

EFFICACY (OUTPUT)

Concept

The x-ray *efficacy* of an x-ray tube is the amount of exposure (in milliroentgens) delivered to a point in the center of the useful x-ray beam at a distance of 1 m from the focal spot for 1 mAs of electrons passing through the tube. The efficacy value

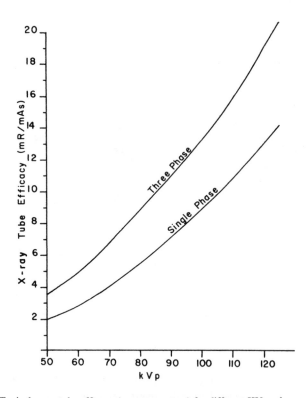

Figure 7-7 Typical x-ray tube efficacy (exposure output) for different KV_p values.

expresses the ability of a tube to convert electronic energy into x-ray exposure. Knowledge of the efficacy value for a given tube permits the determination of patient exposure by methods discussed in a later chapter. The amount of radiation a specific x-ray tube will produce (its efficacy) depends on characteristics of the tube, such as anode material and amount of filtration, the age and condition of the tube, and how the tube is operated. The efficacy, or output, is one of the factors measured as part of the quality control of an x-ray machine.

KV

We have already observed that if the KV is increased, the quantity of radiation (exposure) increases. In other words, the efficacy of x-ray production is very much affected by the KV values we select. Figure 7-7 shows the general relationship between the quantity of radiation produced by a tube (efficacy) and KV_p values.

Notice that it is not a straight line (or linear) relationship. It is normally assumed that the radiation output is proportional to the square of the KV_p. Doubling KV_p increases the exposure from the tube by a factor of about 4 times.

Waveform

Waveform is a characteristic that describes the manner in which the KV changes with time during the x-ray production process; several different KV waveforms are used. The two most common waveform types are *single-phase* and *three-phase*, which will be described in detail in the next chapter. Of these two, three-phase has the least fluctuation and produces more radiation per unit of MAS. In other words, if a single-phase machine and a three-phase machine are set to have the same technique factors (KV and MAS), the three-phase equipment will produce more radiation.

Study Activities

Explain the function of an x-ray tube cathode.

Explain the function of an x-ray anode.

Identify the material the anode is made of.

Explain what happens to electrons as they move from the cathode to the anode.

Explain what happens to electrons when they strike the surface of the anode.

Explain the difference between the electrons that flow into an x-ray tube and those that flow out of the tube.

Describe the Bremsstrahlung process.

Draw the x-ray spectra for two x-ray beams, one produced at 60 KV_p and the other at 120 KV_p.

Describe characteristic x-ray production.

Explain the difference between the spectra for Bremsstrahlung and characteristic x-ray production.

Explain the relationship between x-ray production efficiency and KV.

Use an appropriate graph and determine the exposure at a distance of 1 m produced by a single-phase x-ray machine operating at 90 KV_p and 10 mAs.

Chapter 8

The X-Ray Generator

To produce x-radiation, the x-ray tube must be supplied with electrical energy. The electrical energy provided by a power company is not in the correct form for direct application to the x-ray tube. An x-ray machine has a number of components that rearrange, control, and perhaps store electrical energy before it is applied to the x-ray tube. These components are collectively referred to as either the *power supply* or the *generator*. The function of the generator is not to supply or generate energy but to transform it into an appropriate form for x-ray production. The other major function of the generator is to permit the operator to control three quantities: (1) KV, (2) MA, and (3) exposure time.

The more specific functions of the generator are identified in Figure 8-1. These include

- increasing voltage (produce KV)
- converting AC to DC
- changing waveform
- storing energy (for portable machines)
- controlling KV
- controlling tube current (MA)
- controlling exposure time

A simplified circuit diagram of an x-ray machine is shown in Figure 8-2. We now consider the functions of the components.

KV PRODUCTION

One requirement for x-ray production is that the electrons delivering energy to the x-ray tube must have individual energies at least equal to the energy of the

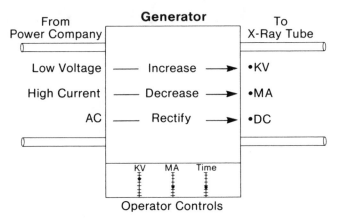

Figure 8-1 Functions performed by an x-ray machine generator.

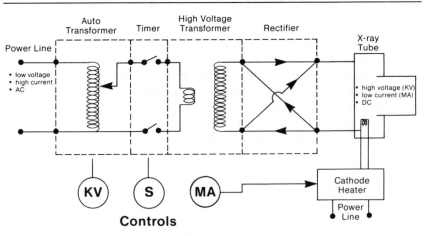

Figure 8-2 A basic circuit of an x-ray machine.

x-ray photons; the x-ray photon energy (in kiloelectronvolts) is always limited by the electron energy, or voltage (in kilovolts).

The electrical energy from a power company is generally delivered at 120, 240, or 440 V. This voltage must be increased to the range of 25,000 to 150,000 V to produce diagnostic-quality x-rays.

Transformer Principles

The device that can increase voltage is the *transformer*, which is one of the major components of the generator. It is a relatively large device connected by

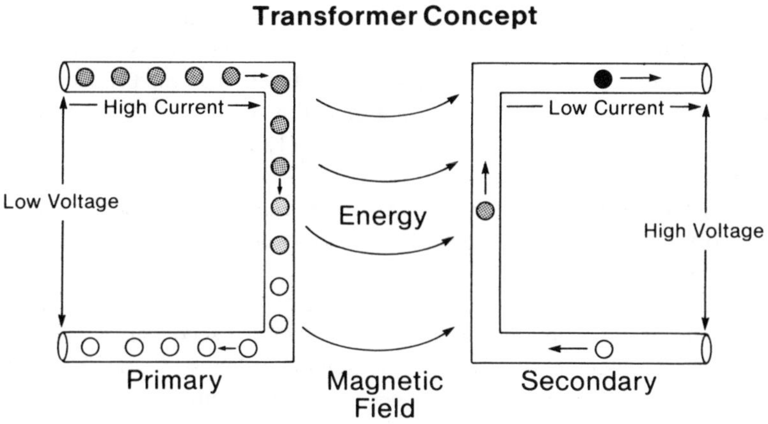

Transformer Concept

Figure 8-3 Electron-energy transfer in a transformer.

cables to the x-ray tube. The basic function of a transformer is illustrated in Figure 8-3.

A transformer has two separate circuits. The input circuit, which receives the electrical energy, is designated the primary circuit, and the output circuit is designated the secondary. Electrons do not flow between the two circuits; rather, energy is passed from the primary circuit to the secondary circuit by a magnetic field.

As electrons flow into the transformer and through the primary circuit, they transfer energy to the electrons in the secondary circuit. The voltage (individual electron energy) increases because the transformer collects the energy from a large number of primary-circuit electrons and concentrates it into a few secondary-circuit electrons. In principle, the transformer repackages the electron energy; the total energy entering and leaving the transformer is essentially the same. It enters in the form of high current, low voltage and leaves in the form of high voltage, low current.

Transformers are designed to produce specific changes in voltage. The transformer described above increases voltage and is therefore designated a *step-up transformer*. For some applications, transformers are designed to decrease voltage and are designated *step-down transformers*.

The High-Voltage Transformer

The high-voltage transformer in most x-ray machines has a voltage step-up ratio of approximately 1,000 to 1. The output of such a transformer would be 1,000 V (1 kV) for each volt applied to the primary circuit.

In a step-up transformer, the current (electron flow) must be larger in the primary circuit than in the secondary. The ratio of the currents is the same as the voltage ratio, except it is reversed. The larger current is in the primary, and the smaller current is in the secondary. For a transformer with a 1,000 to 1 ratio, the current flowing through the primary must be 1 A (1,000 mA) per 1 mA of current flowing through the secondary.

The high-voltage transformer in an x-ray machine can be described in quantitative terms as a device that converts volts into kilovolts and converts amperes into milliamperes.

A transformer physically consists of two coils of wire, as shown in Figure 8-2. One coil forms the primary circuit and the other the secondary circuit of the transformer. Each coil contains a specific number of loops or turns. The characteristic of the transformer that determines the voltage step-up ratio is the ratio of the number of turns (loops) in the secondary coil to the number in the primary. The voltage step-up ratio is determined by, and is the same as, the secondary-to-primary-turns ratio.

There is no direct flow of electrons between the primary and secondary coils; energy is transferred by the magnetic field produced by current passing through the primary coil. The transformer is based on two physical principles involving the interaction between electrons and magnetic fields: (1) When electrons flow through a coil of wire, a magnetic field is created around the coil; (2) electrons within a coil of wire will receive energy if the coil is placed in a magnetic field that is not steady but is pulsing.

The key to transformer operation is that the primary coil must produce a constantly changing, or pulsing, magnetic field to boost the energy of the electrons in the secondary coil. This occurs when the primary coil of the transformer is connected to an AC source. When AC is applied to the input of a transformer, the primary coil produces a pulsing magnetic field. It is this pulsing magnetic field that pumps the electrons through the secondary coil. An electron in the secondary coil gains a specific amount of energy each time it goes around one loop, or turn, on the coil. Therefore, the total energy gained by an electron as it passes through the secondary coil is proportional to the number of turns on the coil. Since the energy of an electron is directly related to voltage, it follows that the output voltage from a transformer is proportional to the number of turns on the secondary coil.

The Autotransformer

In most x-ray apparatuses it is desirable to change the KV applied to the tube to accommodate clinical needs. This is generally done by using the type of transformer illustrated in Figure 8-2, the *autotransformer*, which has a movable contact on the secondary coil that permits the effective number of turns to be changed. Since the output voltage is proportional to the number of turns in the secondary, it can be adjusted by moving the contact. The typical system has an autotransformer that applies an adjustable voltage to the input of the high voltage (step-up) transformer. The autotransformer does not significantly increase voltage; in fact, most slightly reduce voltage. Autotransformers often use a combined primary-secondary coil, but the principle is the same as if the two coils were completely separated.

The question is often raised as to the possibility of putting an adjustable contact on the secondary coil of the high-voltage transformer for the purpose of KV selection. Unfortunately, high voltage, such as that generated in the transformer, requires extensive insulation normally achieved by placing the high-voltage transformer in an enclosed tank of oil. There is no practical way to connect the KV selector switch, located on the control console, to the high-voltage circuit without interfering with the insulation.

RECTIFICATION

The output voltage from the high voltage transformer is AC and changes polarity 60 times per second (60 Hz). If this voltage were applied to an x-ray tube, the anode would be positive with respect to the cathode only one half of the time. During the other half of the voltage cycle, the cathode would be positive and would tend to attract electrons from the anode. Although the anode does not emit electrons unless it is very hot, this reversed voltage is undesirable. The AC from the transformer must be converted to DC before it passes through the x-ray tube. A circuit is needed that will take the voltage during one half of the cycle and reverse its polarity, as illustrated in Figure 8-4. This procedure is called *rectification*.

Rectifiers

The typical rectification circuit is made up of several rectifiers. A rectifier is a relatively simple device that permits electrons to flow through it in one direction but not the other. It can be compared with the valves of the heart, which permit blood to flow in one direction but not the other. In fact, in some countries, rectifiers are referred to as valves. Earlier x-ray equipment used vacuum tube rectifiers, but most rectifiers are now solid-state.

Rectifier Circuits

Notice that the circuit in Figure 8-4 has two input points, to which the incoming voltage from the transformer is applied, and two output points, across which the

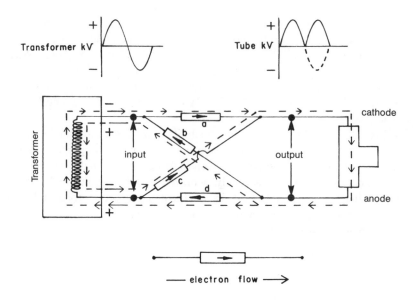

Figure 8-4 A typical full-wave rectifier circuit.

rectified output voltage will appear and be applied to the tube. The circuit contains four rectifiers, labeled *a, b, c,* and *d.* Electrons (current) can flow through a rectifier only in the direction indicated by the arrow. The waveform shown indicates the polarity of the lower points with respect to the upper. The operation of this circuit can be easily understood by considering the following sequence of events.

During the first half of the voltage cycle, the upper transformer conductor is negative, and the electrons flow into the rectifier circuit at that point. From there, they flow only through rectifier *a* and on to the x-ray tube. They enter the tube at the cathode terminal, leave by means of the anode, and return by the lower conductor to the rectifier circuit. At that point, it would appear they have two possible pathways to follow. They flow, however, only through rectifier *d,* because the lower transformer conductor is positive and is more attractive than the upper, negative conductor. During this part of the voltage cycle, rectifiers *b* and *c* do not conduct. During the second half of the cycle, the polarity of the voltage from the transformer is reversed, and the lower conductor is negative. The electrons leave the transformer at this point and pass through rectifier *c* and on to the cathode. Electrons leaving the x-ray tube by means of the lower conductor pass through rectifier *b* because of the attraction of the upper transformer conductor, which is then positive.

Full-Wave

In effect, the rectifier circuit takes an alternating polarity voltage and reverses one half of it so that the voltage coming to the x-ray tube always has the same polarity. In this particular circuit, the cathode of the x-ray tube always receives a negative voltage with respect to the anode. This circuit, consisting of four rectifier elements connected as shown, makes use of all of the voltage waveform, and is therefore classified as a full-wave rectifier.

Half-Wave

A rectifier circuit can have only one rectifier element. The disadvantage is that it conducts during only one half of the cycle. This type is classified as a half-wave rectifier. This type of rectification is found in some smaller x-ray machines, such as those used in dentistry. In such an apparatus, the x-ray tube itself often serves as the rectifier.

VOLTAGE WAVEFORM AND X-RAY PRODUCTION

Single-Phase

Both the full-wave and half-wave circuits discussed up to this point use single high-voltage transformers and are classified as *single-phase apparatuses*. The basic disadvantage of single-phase operation is that the KV applied to the x-ray tube constantly changes, or pulses, throughout the exposure, as shown in Figure 8-5. This means that both the quantity and energy spectrum of the x-rays produced change constantly with time throughout the cycle. The output from the tube is a spectrum of photon energies that is an average of the different spectra produced at each instant throughout the KV cycle.

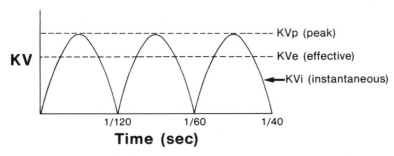

Figure 8-5 Relationship of KV peak, effective, and instantaneous values for a single-phase generator.

Three principal KV values are associated with the typical single-phase wave-form. Each is related to an aspect of x-ray production. At any instant in time, the pulsing KV has an *instantaneous value* (KV$_i$), which determines the rate of x-ray production at each specific instant. During each cycle, the KV reaches a maximum or *peak value* (KV$_p$). It is the KV$_p$ that is set by the operator as a control on x-ray production. The ability of a voltage to transfer energy is related to its *effective value* (KV$_e$), which reflects the fact that the voltage varies with time and does not always produce energy at the peak value. For the typical sine-waveform voltage, as shown in Figure 8-5, the KV$_e$ is 70.7% of the KV$_p$. Our primary interest in the KV$_e$ is that it determines the rate at which heat is produced in the x-ray tube.

Constant Potential

Some x-ray generators produce a constant KV; in these cases the KV$_p$, KV$_e$, and KV$_i$ have the same value. These generators are called *constant potential* equipment. The constant potential x-ray machine produces more photons with higher average or effective energy than are produced by the single-phase machine, as shown in Figure 8-6.

The rate at which exposure is delivered to the receptor varies significantly with time for single-phase equipment, as shown in Figure 8-6. Most of the exposure is produced during a small portion of the voltage cycle, when the voltage is near the KV$_p$ value. Several factors contribute to this effect. One is that the efficiency of x-ray production increases with voltage and gives more exposure per milliampere-second at the higher voltage levels. Second, the photons produced at the higher tube voltages have higher average energies and are more penetrating. Third, the MA also changes with time during the voltage cycle.

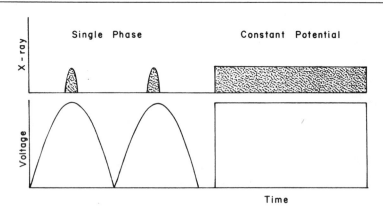

Figure 8-6 Comparison of single-phase and constant potential x-ray production.

When an x-ray machine is set at a certain MA value, the stated value is usually the average throughout the exposure time. In single-phase equipment, the MA value changes significantly during the voltage cycle. The effect is that the x-ray exposure is delivered to the receptor in a series of pulses. Between the pulses is a period of time during which no significant exposure is delivered. This means, generally, that the total exposure time must be longer for single-phase than for constant potential x-ray equipment, which can deliver a given film exposure in a much shorter total exposure time.

Three-Phase

One of the most practical means of obtaining essentially constant voltage and high average current is to use *three-phase electrical power*. The concept of three-phase electricity can best be understood by considering it as three separate incoming power circuits, as shown in Figure 8-7. Although this illustration shows six conductors coming in, this is not necessary in reality because the power lines can be shared by the circuits. Each circuit, or phase, delivers a voltage that can be transformed and rectified in the conventional manner. The important characteristic of a three-phase power system is that the waveforms or cycles in one circuit are out of step (out of phase), with those in the other two. This means that the voltage in the three circuits peaks at different times. In an actual circuit, the three voltage

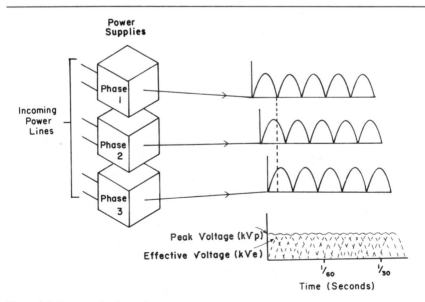

Figure 8-7 Concept of a three-phase generator.

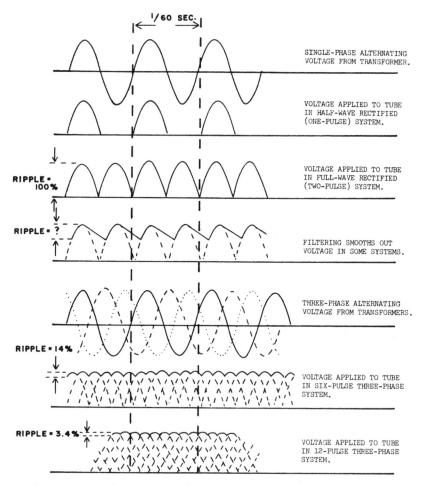

Figure 8-8 Voltage waveforms encountered in x-ray production.

waveforms are combined, as shown in Figure 8-7. They are not added but are combined so that the output voltage at any instant is equal to that of the highest phase at the time. Since the voltage drops only a few percent before it is picked up by another phase, the KV_i at all times is quite close to the KV_p.

The voltage variation over the period of a cycle is designated the *ripple* and is expressed as a percentage. The typical ripple levels for several power supply types are shown in Figure 8-8. One way to classify power supply circuits is according to the number of pulses they produce in the period of one cycle; for a frequency of 60 hertz this is 1/60th of a second. By using a complex circuit of transformers and rectifiers, it is possible to produce a 12-pulse machine that has a ripple level of less than 4%.

CAPACITORS

Capacitors (sometimes referred to as condensers) are electrical components used in many types of electronic equipment. A capacitor consists of two electrical conductors, such as sheets of metal foil, separated by a layer of insulation. They are used in some x-ray generators for two purposes. In capacitor discharge (or condenser discharge) portable x-ray machines, capacitors are used to accumulate and store electrical energy; in other types of equipment they are used as a filtering device to produce constant potential KV.

Capacitor Principles

The basic function of a capacitor is illustrated in Figure 8-9. A capacitor can be described as a storage tank for electrons. When it is connected to a voltage source, electrons flow into the capacitor, and it becomes "charged." As the electrons flow in, the voltage of the capacitor increases until it reaches the voltage of the supply. Energy is actually stored in the capacitor when it is charged; the amount stored is proportional to the voltage and the quantity of stored electrons. If a charged capacitor is connected to another circuit, the capacitor becomes the source, and the electrons flow out and into the circuit.

Energy Storage

In the discussion of the high-voltage transformer, it was pointed out that the current flowing into the power supply circuit must be greater than the tube current

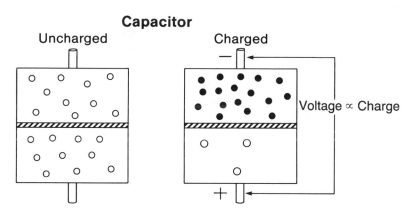

Figure 8-9 Concept of a capacitor.

by a factor equal to the voltage step-up ratio. This is typically as high as 1,000 to 1, which would require 1 A of power line current for every 1 mA of tube current. The special circuits feeding permanently installed x-ray equipment can accommodate these high currents. In other areas of the hospital where portable x-ray equipment is used, the normal electrical circuits are limited to currents of about 15 to 20 A. Many portable x-ray machines are now designed to overcome this limitation by using a capacitor as an electron-energy storage device.

A simplified capacitor storage-discharge power supply is shown in Figure 8-10. Step-up transformers and rectifiers are used to produce the high voltage and to charge the capacitor. The electrons are pumped into the capacitor. Charging times can be as long as 10 to 20 seconds. The current flow into the capacitor is typically only a few milliamperes; when it is discharged to the tube over a short period of time (ie, during the exposure time), the current can be several hundred milliamperes.

The voltage across a capacitor is proportional to the quantity of electrons stored (MAS); the actual relationship depends on the size, or capacity, of the capacitor. Many machines use 1-microfarad (μF) capacitors, which produce a voltage of 1 kV for each milliampere-second stored. As the electrons flow from the capacitor to the tube, the voltage drops at the rate of 1 kV/mAs. For example, if a machine is charged to 70 kV and an exposure of 18 mAs is made, the voltage will have dropped to 52 kV at the end of the exposure.

Capacitor-storage x-ray equipment has a high-voltage waveform, unlike other power supplies. An attempt to obtain large milliampere-second exposures drops the kilovolts to very low values by the time the exposure terminates. Since low tube voltages produce very little film exposure but increase patient exposure, this

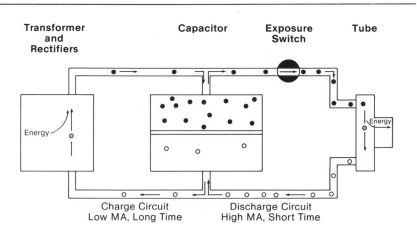

Figure 8-10 Basic capacitor-storage x-ray machine.

type of operation should be avoided. The total MAS should generally be limited to approximately one third of the initial KV value.

A means for turning the tube current on and off is included in the x-ray tube circuit. Most machines use a grid-control x-ray tube for this purpose.

Filtration

When a capacitor is used to produce a constant voltage, or potential, it is permanently connected between the rectifier circuit and the x-ray tube. As the voltage rises toward its peak, electrons from the rectifier circuit flow both to the x-ray tube and into the capacitor. When the voltage from the rectifier circuit begins to fall, electrons flow out of the capacitor and into the x-ray tube. Within certain operational limits, this can maintain a constant potential across the tube. Capacitors can be used on single-phase and three-phase equipment. On single-phase equipment, the maximum MA that can be used without introducing ripples is more limited.

HIGH-FREQUENCY POWER SUPPLIES

Another approach used for some machines producing relatively constant KV is to convert the 60-Hz (low-frequency) electricity to a higher frequency before it is rectified. This function is performed by an electrical circuit known as an *invertor*. After the high-frequency voltage is rectified, the short-duration pulses are much easier to filter into an essentially constant potential.

MA CONTROL

The cathode is heated electrically by a current from a separate low-voltage power supply. The output of this supply is controlled by the MA selector. Increasing the MA selector setting passes more heating current through the cathode; this, in turn, increases the temperature, and the increased emission produces an increase in x-ray tube current. There are actually two currents flowing through portions of the x-ray tube: one, the MA, flows from the cathode to the anode and through the high-voltage power supply; the other flows only through the filament of the cathode. It is this second current that controls the cathode-to-anode current.

The cathode temperature required to produce adequate thermionic emission, especially at high MA values, is relatively high. The temperature is sufficiently high, in many cases, to produce some evaporation of the tungsten cathode.

Because of this, it is undesirable to keep the cathode at the high operating temperature except for the duration of the x-ray exposure. Most x-ray equipment operates with two levels of cathode heating. When the equipment is turned on, the cathode is heated to a standby level that should not produce significant evaporation. Just before the actual exposure is initiated, the cathode temperature is raised to a value that will give the appropriate tube current. In most radiographic equipment, this function is controlled by the same switch that activates the anode rotor. Unnecessarily maintaining the cathode temperature at full operating temperature can significantly shorten the x-ray tube's lifetime.

The x-ray tube current can be read or monitored by a meter located in the high-voltage circuit; it must be placed in the part of the circuit that is near ground voltage, or potential. This permits the meter to be located on the control console without extensive high-voltage insulation.

EXPOSURE TIMING

Another function of the generator is to control the duration of the x-ray exposure. In radiography the exposure is initiated by the equipment operator and then terminated either after a preset time has elapsed or when the receptor has received a specific level of exposure. In fluoroscopy the exposure is initiated and terminated by the operator, but a timer displays accumulated exposure time and produces an audible signal at the end of each 5-minute exposure increment.

Operator-controlled switches and timers turn the radiation on and off by activating switching devices in the primary circuit of the x-ray generator.

Manual Timers

X-ray equipment with *manual timers* requires the operator to set the exposure time before initiating the exposure. The time is determined by personal knowledge or from a technique chart, after the size of the patient and the KV and MA values being used are considered.

Exposure timers for single-phase equipment usually operate in increments of 1/120th of a second, which is the duration of one half of a 60-Hz voltage cycle. This is also the elapsed time between individual pulses of radiation. It is generally not practical to terminate an exposure during the actual radiation pulse.

A potential problem with this type of timer is its inability to make relatively small adjustments in exposure time. In a situation in which 1/120th of a second produces a slight underexposure, the next possible exposure value, 1/60th of a second, would double the amount of radiation and probably result in an overexposed film. This problem is especially significant for the shorter exposure times.

Three-phase and constant potential equipment produces radiation at a more constant rate, and the exposures can be timed and terminated with more precision.

Automatic Exposure Control

Automatic exposure control (AEC) is an x-ray machine function that terminates the exposure when a specific predetermined amount of radiation reaches the receptor. This function is also referred to as *phototiming*. AEC is used frequently in many general radiographic procedures and is always used in spot filming and cineradiography.

QUALITY ASSURANCE PROCEDURES

With all x-ray equipment, the operator can control the quantity (exposure) and quality (penetrating ability) of the radiation with the KV, MA, and exposure-time controls. If the equipment is not properly calibrated or is subject to periodic malfunction, it will not be possible to control the radiation output. This can result in reduced image quality and unnecessary patient exposure, especially when repeat images are required.

X-ray equipment is required to meet certain federal standards at the time of installation, and in most states periodic calibration and quality assurance inspections are required.

Study Activities

Name the major components that make up an x-ray generator.

Explain the primary functions of a generator.

Explain how energy passes from the primary to the secondary circuits of a transformer.

Explain how a transformer produces a high voltage.

Explain how the KV is changed in an x-ray machine.

Explain how the MA is controlled in an x-ray machine.

Compare the voltage in the primary and secondary circuits of a high-voltage transformer.

Compare the currents in the primary and secondary circuits of a high-voltage transformer.

Explain how current flows through a rectifier.

Explain the difference between a full-wave and a half-wave rectifier.

Draw the KV waveform produced by a full-wave rectifier circuit.

Use your drawing to show the difference between KV_p and KV_e

Explain why constant potential is generally better than single phase for x-ray production.

Describe the KV waveform produced by a three-phase generator.

Identify the waveform that produces the most ripple.

Identify the waveform that produces the least ripple.

Explain how a capacitor stores electrical energy.

Chapter 9

Interaction of X-Radiation with Matter

We have seen that x-ray photons are created when energetic electrons flying through the x-ray tube interact with the anode and lose their energy. X-ray photons end their lives by transferring their energy back to electrons contained in various materials. X-ray interactions are important in radiography for many reasons. For example, it is the interaction and absorption of x-ray photons with the structures of the human body that produce the x-ray image; the interaction of photons with the receptor converts the x-ray image into one that can be viewed or recorded. The interaction of x-ray photons and human tissue can produce undesirable biological effects.

This chapter considers the basic interactions between x-ray photons and matter.

INTERACTION TYPES

Recall that photons are individual units of energy. As an x-ray beam passes through an object, three possible fates await each photon, as shown in Figure 9-1:

1. It can penetrate the object without interacting.
2. It can interact and be completely absorbed by depositing its energy.
3. It can interact and be scattered or deflected from its original direction and deposit part of its energy.

There are two kinds of interactions through which photons deposit their energy; both are with electrons. In one type of interaction the photon loses all its energy; in the other it loses a portion of its energy, and the remaining energy is scattered. These two interactions are shown in Figure 9-2.

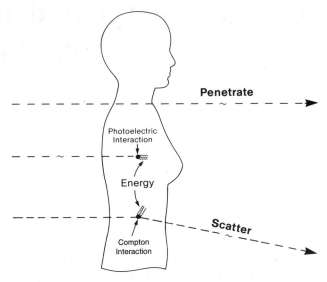

Figure 9-1 Photons entering the human body will either penetrate, be absorbed, or produce scattered radiation.

Photoelectric

In the *photoelectric* (photon-electron) *interaction*, as shown in Figure 9-2, a photon transfers all its energy to an electron located in one of an atom's shells. The electron is ejected from the atom by this interaction and begins to pass through the surrounding matter. The electron rapidly transfers its energy to the matter and moves only a relatively short distance from its original location. The photon's energy is therefore absorbed in the matter close to the site of the photoelectric interactions. The photoelectric interaction is the first step of a two-step process that deposits a photon's energy in matter.

Photoelectric interactions usually occur with electrons that are firmly bound to the atom, that is, those with a relatively high binding energy. Electrons with high binding energies are found most frequently in the chemical elements with relatively high atomic numbers. Calcium, iodine, and barium produce many more photoelectric interactions than the basic elements in soft tissue, hydrogen, oxygen, carbon, and nitrogen.

Compton

A *Compton interaction* is one in which only a portion of the photon energy is transferred to the electron and absorbed in the matter. The remaining photon with reduced energy leaves the site of the interaction in a direction different from that of

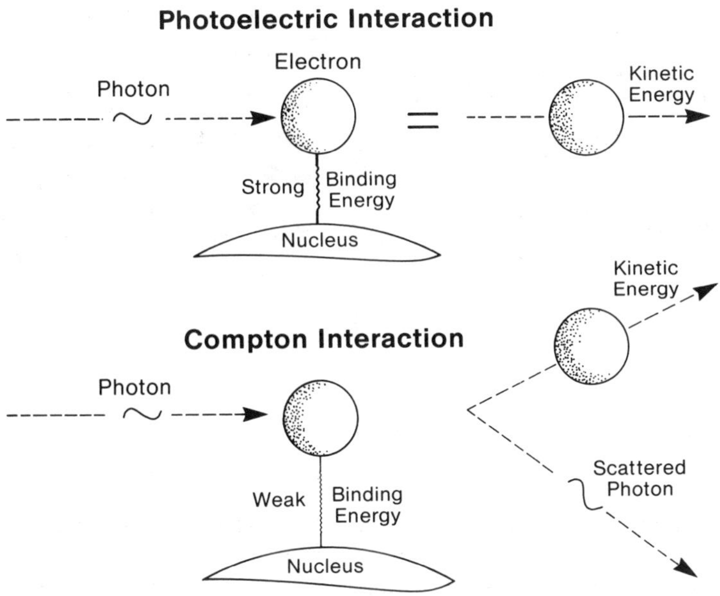

Photoelectric Interaction

Compton Interaction

Figure 9-2 The two basic interactions between photons and electrons.

the original photon, as shown in Figure 9-2. Because of the change in photon direction, this type of interaction is classified as a *scattering* process. This is significant in some situations because the materials within the primary x-ray beam become a secondary radiation source. The most significant object producing scattered radiation in an x-ray procedure is the patient's body. The portion of the patient's body that is within the primary x-ray beam becomes the actual source of scattered radiation. This has two undesirable consequences. The scattered radiation that continues in the forward direction and reaches the image receptor decreases the quality (contrast) of the image; the radiation that is scattered from the patient is the predominant source of radiation exposure to the personnel conducting the examination.

Compton interactions occur most frequently with electrons that have relatively weak binding energies. Virtually all of the electrons in the chemical elements that make up soft tissue have weak binding energies. Therefore, Compton interactions that produce scattered radiation are especially significant in soft tissue.

Electron Interactions

We have just seen that both photoelectric and Compton interactions transfer energy from x-ray photons to electrons. These electrons are propelled from the site

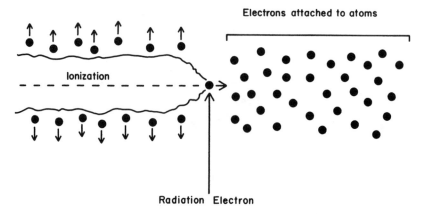

Figure 9-3 Ionization produced by a radiation electron.

of the interaction at a high velocity. They are carrying the energy taken from the photons in the form of kinetic energy or energy of motion.

As the electrons leave the interaction site, they immediately begin to transfer their energy to the surrounding material, as shown in Figure 9-3. Because the electron carries an electric charge, it can interact with other electrons without touching them. As it passes through the material, the electron in effect pushes the other electrons away from its path. If the force on an electron is sufficient to remove it from its atom, ionization results. X-radiation is classified as a form of ionizing radiation because when x-ray photons interact with various forms of matter, in either photoelectric or Compton interactions, the end result is ionization.

The electrons carrying the energy away from a photoelectric or Compton interaction travel a very short distance (a fraction of a millimeter) in materials such as human tissue. This means that the energy from the x-ray photon is absorbed in the body very close to the site of the interaction.

PHOTON INTERACTION RATES

Attenuation

As a photon makes its way through matter, there is no way to predict precisely either how far it will travel before engaging in an interaction or the type of interaction it will engage in. In radiography we are not concerned with the fate of an individual photon but rather with the collective interactions of the large number of photons that make up the x-ray beam.

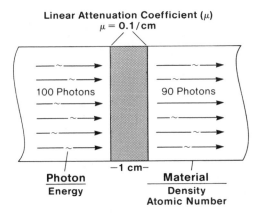

Figure 9-4 Linear attenuation coefficient.

Let us observe what happens when a group of photons encounters a slice of material that is 1 cm thick, as illustrated in Figure 9-4. Some of the photons interact with the material, and some pass on through. The interactions, either photoelectric or Compton, remove some of the photons from the beam. This is known as *attenuation*. Under specific conditions, a certain percentage of the photons will interact, or be attenuated, in a one-unit thickness of material. The actual fraction of photons interacting per one-unit thickness of material is designated the *linear attenuation coefficient* (μ). In our example the fraction that interacts in the 1-cm thickness is 0.1, or 10%, and the value of the linear attenuation coefficient is therefore 0.1 per cm.

The value of the linear attenuation coefficient indicates the rate at which photons interact as they move through material. This is determined by the energy of the individual photons and the atomic number and density of the material.

If conditions are such that very few photons are interacting, the attenuation coefficient will have a low value. On the other hand, if conditions are producing a high rate of interactions (either photoelectric or Compton), the attenuation coefficient will have a high value. Let us now consider factors or conditions that can change the rate of photon interactions and the attenuation.

Photon Energy

In a given material, some photons are quicker to interact than others. This is especially true for photoelectric interactions. Figure 9-5 shows the relationship between the rate of interactions (attenuation coefficient) between photons and the electrons in iodine. This particular interaction is important in radiography because

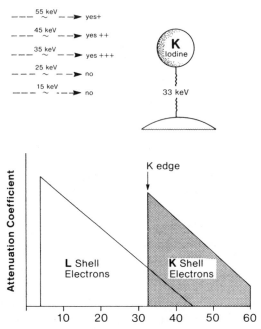

Figure 9-5 Relationship between the probability of photoelectric interactions and photon energy.

iodine is used as a contrast media for most vascular examinations. A high interaction rate (attenuation coefficient value) is desirable for this sort of examination because it produces better image contrast. However, the point we are considering here is that the rate of interactions in iodine depends on the energy of the individual photons.

In the iodine atom there are two groups of electrons that engage in photoelectric interaction: the K-shell group and the L-shell group. The basic difference between these two groups of electrons is the energy with which they are bound to the atom. The K-shell electrons have a binding energy of 33 keV, and the L-shell electrons have much lower binding energies. It is the interaction with the K-shell electrons that produces most of the image contrast when iodine is used as contrast medium.

One of the requirements for a photoelectric interaction is that the x-ray photon must have an energy that is at least equal to the binding energy of the electron. The photons with the best chance of interacting are those with an energy that is just a little more than the energy of the electron. If a photon has an energy that is much greater than the binding energy of the electron, it can interact, but the number of interactions will be relatively few. Now let us examine the group of photons in Figure 9-5 as they encounter a K-shell electron in iodine. The two photons with energies below 33 keV cannot interact with the K-shell electrons; they can, however, interact with the L-shell electrons. The 35 keV photon will have a very good chance (yes + + +) of interacting, because it has an energy that is just a little greater than the electron binding energy (33 keV). This level of photon energy is highly desirable in angiography because it produces excellent image contrast. The photons with higher energies than this can interact, but the number that do decreases as the energy is increased.

The graph shows two significant features of the relationship between the number of interactions (attenuation coefficient) and photon energy. One is that the coefficient value, or the probability of photoelectric interactions, decreases rapidly with increased photon energy. This means, in general, that high-energy photons have a smaller chance of interacting (being attenuated) than low-energy photons. Photons that are not attenuated are the ones that penetrate through a specific material. Therefore, high-energy photons are usually better penetrators than low-energy photons.

The other important feature of the attenuation coefficient-photon energy relationship is the very abrupt change that occurs at 33 keV, the binding energy of the K-shell electrons in iodine. When the photons have energies below this value, there is a relatively low rate of interaction because they can interact only with the L-shell electrons. However, when photons have energies above this value and can interact with the K-shell electrons, the attenuation rate jumps up rather dramatically. This feature of the attenuation coefficient graph is known as the *K edge*.

The K edge is the dividing line between low interaction rate (on the left) and high interaction rates (on the right) as shown in Figure 9-5. The energy at which the K edge occurs is a characteristic of each chemical element. We have used iodine in this example. Other elements will have different K edge energies.

We will come back to this point later when we discuss how iodine contrast can be changed by changing the KV. Think about it for a moment; when you change the KV, you are changing the energy of the photons in the x-ray beam. This, in turn, is changing their chance for interacting with iodine atoms, and that changes image contrast.

Compton interactions can occur with the very loosely bound electrons in atoms. All electrons in materials with low atomic numbers and the majority of electrons in materials with high atomic numbers are in this category.

The rate at which Compton interactions occur does not depend on photon energy to any significant degree. This means that photons throughout the energy spectrum engage in Compton interactions.

SUMMARY

As x-ray photons pass through matter, they can engage in either photoelectric or Compton interactions with the material's electrons. The photoelectric interaction captures all photon energy and deposits it within the material, whereas the Compton interaction removes only a portion of the energy, and the remainder continues as scattered radiation. The combination of the two types of interactions produces the overall attenuation of the x-ray beam.

We now consider the factors that determine which of the two interactions is most likely to occur in a given situation.

The rate at which interactions occur can depend on the energy of the x-ray photons and the density and atomic number of the material. The rate of photoelectric interactions changes rather dramatically with photon energy. Photoelectric interactions also occur most frequently in materials with relatively high atomic numbers. Compton interactions occur at a relatively constant rate for most materials found in the human body and at the photon energies found in the typical x-ray beam.

In soft tissue, photoelectric interactions occur with a relatively high rate and are the most significant type of interaction for photons with energies less than approximately 30 keV. Photons with energies greater than this will most frequently engage in Compton interactions.

Materials that have a relatively high atomic number (calcium, iodine, barium, lead, etc.) will generally produce a relatively high rate of photoelectric interactions.

The rates of all types of interactions (and therefore the rate of attenuation) are proportional to the physical density of the material. We see this effect in radiographs of the chest, where the low-density lungs absorb and attenuate less than the other, more dense parts of the body.

Study Activities

Explain what happens to an x-ray photon in a photoelectric interaction.

Explain what happens to an x-ray photon in a Compton interaction.

Identify the electrons within an atom that most frequently engage in photoelectric interactions.

Explain what happens to an electron that is involved in a photoelectric interaction.

Identify the photon energy that has the highest rate of absorption in iodine.

Identify the iodine electron shells that a 30-keV photon can interact with.

Define attenuation coefficient.

Identify the most common type of interaction between 50-keV photons and soft tissue.

Identify the most common type of interaction between 50-keV photons and iodine contrast media.

X-Ray Penetration

The reason we can use x-radiation to produce an image of the internal anatomy is because this type of radiation can penetrate the human body. This is one way in which x-rays are different from light. Light will not pass through the thicker parts of the body, but x-rays will. However, it is important to observe that an x-ray beam does not completely penetrate most objects. Generally speaking, if we apply an x-ray beam to an object, such as the human body, some fraction of the x-ray beam photons will pass through (penetrate) it. The others will either be absorbed in the object or will scatter off in some other direction.

The amount (fraction) of radiation that penetrates an object depends on certain characteristics of both the object and the radiation. We will soon learn what these characteristics are and how we can use them to control x-ray penetration in radiography.

There are times when we want a relatively high penetration. One example is in chest radiography, where it is usually desirable to use a very penetrating x-ray beam in order to reduce area contrast. Also, an x-ray beam with high penetration actually reduces the exposure to the patient because more of the radiation passes through to the receptor rather than being absorbed in the tissue.

There are also times when low penetration is desirable. We usually get better contrast of objects in the body, such as blood vessels and calcifications, when the x-rays' penetration through them is relatively low. Also, we would like for any penetration through radiation shields such as lead aprons to be as low as possible.

Let's now look at Figure 10-1 to get a better picture of penetration. We remember that an x-ray beam is made up of many small individual photons. When an x-ray beam is passing through an object, all photons do not travel the same distance before they have an interaction. Some will interact very close to where they enter the object, and others will travel a greater distance before interacting. Of course, some photons will pass through the object without interacting at all. These are the ones that penetrate. The fraction of the original photons that penetrate

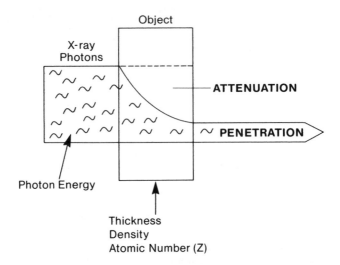

Figure 10-1 Factors that affect the penetration of radiation through a specific object.

depends on characteristics of both the photons and the object through which they are passing.

The characteristic of x-radiation that affects its penetrating ability is the energy of the individual x-ray photons. In materials such as soft tissue, high-energy photons are better penetrators than low-energy photons.

An object has three characteristics that affect penetration: (1) thickness, (2) density, and (3) atomic number. Penetration decreases as thickness is increased. We get less penetration through thick objects than through thin objects. The physical density of an object affects penetration in the same way as thickness. We get less penetration through dense objects (bone, metal, etc) than we do through less dense materials such as air. We have already discovered that materials with atomic numbers within a certain range are very good x-ray absorbers (attenuators). They therefore have reduced penetration. The penetration through iodine and barium contrast media, for example, is relatively low because of these substance's position on the atomic number scale.

HALF-VALUE LAYER

Half-value layer (HVL) is the most frequently used factor for describing the penetrating ability of x-radiation. HVL is the thickness of any given material through which one half of a given amount of x-radiation penetrates. It is expressed in units of distance (mm or cm).

HVL tells us something about how far x-radiation penetrates through a specific material. X-radiation with an HVL of 5 mm of aluminum is much more penetrating than radiation that has an HVL of 2 mm of aluminum.

Let's take a moment to compare x-radiation to cross-country runners. We have two groups, the blue team and the red team. If we watch them run, we might discover that only one half of the blue team can run 2 miles, but one half of the red team members can run 5 miles. It's easy to see that the red team, which has an "HVL" of 5 miles, can go a greater distance (penetrate further) than the blue team, which has an "HVL" of only 2 miles.

Let's now use Figure 10-2 to watch an x-ray beam pass through an object that has been sliced into 1-cm layers. We will start out with 1,000 photons and then observe the number that penetrate through different thicknesses of the material.

One very important characteristic of x-radiation is that a specific thickness of a given material (1 cm) will always absorb the same fraction, but not the same number, of the photons that enter it. The actual fraction absorbed in a 1 cm thickness depends on the energy of the photons and the density and atomic number of the material. In Figure 10-2 we are assuming that the conditions produce an absorption of 1/10th of the protons in each centimeter thickness. As the x-ray beam passes through this material, each slice will absorb 1/10th of the photons that enter it.

As the x-ray beam passes through the material, let's keep score of how many photons are absorbed in each layer and place the number above each slice. Remember, each slice absorbs the same fraction of photons but not the same number. If 1,000 photons enter the first slice, 1/10th of these, or 100 photons, would be absorbed. This leaves only 900 photons to enter the second slice. Therefore, 1/10th of 900 would be 90 photons absorbed in the second slice. 810 photons would enter the third slice, where 81 would be absorbed. This process continues as the x-ray beam makes its way through the material. We notice that the number of photons in the x-ray beam is decreasing as it moves through the material.

Let's now go back and look at how many photons penetrate different thicknesses of the material. The best way to see this is by means of the graph that is drawn directly below the slices in Figure 10-2. If we start with 1,000 photons entering the material (zero thickness) and then subtract the number absorbed in each layer, we will find how many are penetrating through various thicknesses. The numbers would look like this:

Thickness	Photons
0 cm	1,000
1	1,000 − 100 = 900
2	900 − 90 = 810
3	810 − 81 = 729
4	729 − 73 = 656

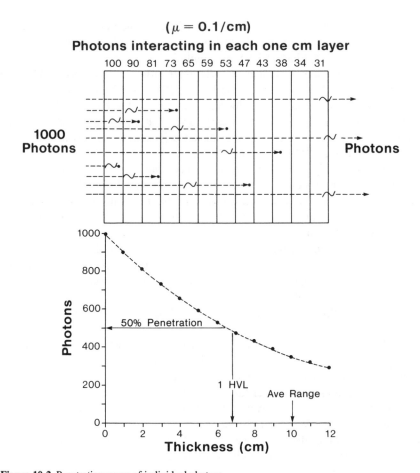

Figure 10-2 Penetration range of individual photons.

If we continue this process, we can find how many photons penetrate through various thicknesses and use the numbers to plot the graph shown in Figure 10-2. As the thickness increases, the fraction of the original 1,000 photons passing through that thickness decreases. We can now use this graph to find the actual value for the HVL. We first go to the point on the graph that represents 50% penetration (500 photons). From this point, we draw a line down to the thickness scale, to find that a thickness of 6.7 cm is the HVL. In other words, for this particular set of conditions, 50% of the x-ray beam photons can penetrate through a thickness of 6.7 cm of material.

The actual thickness of 1 HVL depends on the individual energy of the photons and characteristics of the material.

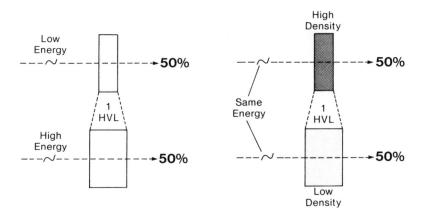

Figure 10-3 Factors that affect the thickness of 1 HVL.

Figure 10-3 illustrates two factors. One HVL is thicker for high-energy photons than for low-energy photons. This is because the high-energy photons are better penetrators, and it takes a greater thickness to stop one-half of them. We also see the effect of density on HVL. It does not require as much high-density material to be 1 HVL.

CHARACTERISTICS OF MATERIAL

We have already identified three characteristics of an object that will have an effect on x-ray penetration through it: thickness, density, atomic number. We will now look at these in more detail and see how each affects the practice of radiography.

Thickness

When considering penetration, we sometimes express the thickness of objects in HVLs. This is a very useful approach because if we know thickness in HVLs, we can very easily determine the penetration, because each HVL thickness reduces the penetration by one half. Let's go to Figure 10-4 to see how this works. The penetration through one HVL is always 50%. That's the definition of an HVL. If we have an object that is two HVLs thick, the penetration will be one half of 50%, or 25%. This is because the second HVL thickness absorbs 50% of the radiation that penetrated through the first HVL. If we now add one additional HVL

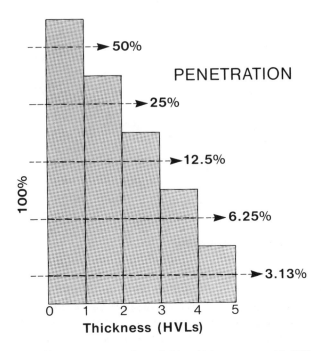

Figure 10-4 Relationship between penetration and object thickness expressed in HVLs.

thickness, for a total of 3 HVLs, the penetration will again be reduced by a half, or down to 12.5%

What we are observing is that penetration decreases as the thickness of an object increases. In theory, we would never arrive at a thickness where the penetration is reduced completely to zero. In practice, there is always some thickness that will reduce the radiation to some very low value.

One practical application of this is that x-ray penetration is much less through thick patients than through thin patients.

Density

In Chapter 6 we learned that different materials have different physical densities. Density is one of the factors that affects penetration. In fact, penetration is inversely related to density. We get much higher penetration through low-density materials than through high-density materials. Density is one of the most important characteristics of tissue influencing the radiographic image. A radiograph of the chest is a good example. Most of the contrast is created by differences in density. Much more radiation penetrates the low-density lungs as compared to the more dense mediastinum.

Radiation shields are usually constructed of high-density materials such as lead and concrete in order to reduce penetration.

Atomic Number

The atomic number of a material will have an effect on penetration. However, the relationship is much more complicated than for thickness and density. The degree to which atomic number affects penetration depends on the energy of the x-ray photons. Penetration is reduced when the rate of photoelectric interactions is high. In Chapter 9 we discovered that this can happen when there is the correct match between the photon energies and the binding energy of the electrons within the material. We should also recall that the atomic number of a material determines its binding energy. Therefore, for a specific x-ray beam spectrum, there are certain chemical elements whose binding energies are a good match, producing high absorption and low penetration. This is why iodine and barium are good contrast agents and produce high absorption and low penetration at certain photon energies.

X-RAY BEAM QUALITY

The general term *quality* refers to an x-ray beam's penetrating ability. It has already been shown that for a given material, the penetrating ability of an x-ray beam depends on the energy of its photons. Up to this point, the discussion has related penetration to one specific photon energy. For x-ray beams that contain a range or spectrum of photon energies, the penetration rate is different for each photon energy. The overall penetration generally corresponds to the penetration achieved by photons with an energy that is somewhere between the minimum and maximum energies of the spectrum of photons in the beam. This energy is designated the *effective energy* of the x-ray spectrum. Therefore, we can change the penetrating ability of an x-ray beam by changing its effective energy. X-ray beam quality is affected by four factors: the x-ray tube anode material, filtration, KV waveform, and KV_p. Most x-ray tubes use tungsten as the anode material, and that cannot be changed by the operator. Each piece of x-ray equipment is designed with a specific waveform (single phase, three-phase, etc), and that cannot be changed by the operator either.

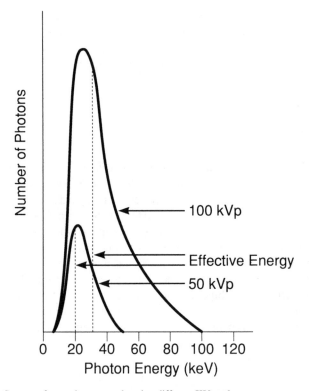

Figure 10-5 Spectra of x-ray beams produced at different KV_p values.

KV_p

The only factor available to the radiographer for controlling x-ray beam penetration is the KV_p. Figure 10-5 compares two x-ray spectra with different KV_p values. When the KV_p is increased from 50 to 100 kV_p, the spectrum contains more high-energy photons. This causes the x-ray beam to have a higher effective energy and to be more penetrating in materials such as soft tissue. The 100 kV_p beam will have a thicker HVL than the 50 kV_p beam.

Filtration

If an x-ray beam that contains many low-energy photons is applied to a patient's body, these photons have a very low chance of penetrating the body and contributing to the formation of the image. Most of these low-energy photons will be absorbed in the tissue and will only increase the radiation dose to the patient.

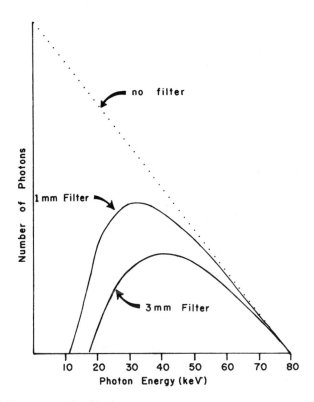

Figure 10-6 X-ray spectra after filtration.

Therefore, it is not desirable for an x-ray beam to contain many low-energy, nonpenetrating photons. Unfortunately, x-ray tubes produce a spectrum that contains a large quantity of low-energy photons. This is the unfiltered spectrum, shown in Figure 10-6. If an x-ray beam with such a spectrum is used, it will expose the patient to a large amount of unnecessary radiation.

Filtration is the process that removes the low-energy photons from the x-ray beam before it reaches the patient's body. All x-ray machines contain a filter. The filter is a collection of material that the x-ray beam passes through as it comes out of the x-ray tube. The filters in most x-ray tubes consist of two parts. One part is the x-ray tube window material, the alignment mirror, and other parts of the tube assembly through which the x-ray beam must pass. This is known as the *inherent filtration*. The inherent filtration is usually not enough, so it must be supplemented by *added filtration*. The added filtration is in the form of thin sheets of metal placed in the pathway of the x-ray beam. Aluminum is the most common material used in x-ray beam filters. However, some mammography machines use molybdenum filters, and some chest radiography machines might use filters of brass or copper.

Table 10-1 Recommended Minimum Penetration (HVL) for Various KV$_p$ Values

KV$_p$	Minimum Penetration (HVL) for Aluminum (mm)
30	0.3
50	1.2
70	1.5
90	2.5
110	3.0

The function of the filter is to absorb the low-energy photons before they reach the patient's body. After an x-ray beam passes through the filter, it will have a very different spectrum, as shown in Figure 10-6. The low-energy, low-penetration photons will be removed by the filter. Most of the higher energy photons will not be affected.

All x-ray machines are required by law to have filtration. The requirement specifies the amount of total filtration (inherent plus added) in millimeters of aluminum.

It can be determined if an x-ray machine has adequate filtration by measuring the HVL of the x-ray beam. This works because adding filtration increases the penetrating ability (HVL) of the x-ray beam, because it removes low-energy photons. The photons remaining in the beam are better penetrators. When physicists calibrate an x-ray machine, one of the factors they measure is the HVL. If the HVL has a certain value at specific KV settings, it is assumed that the machine contains adequate filtration.

Table 10-1 shows the minimum HVL that a radiographic machine should have for different KV values. Since most machines can be operated at up to 110 KV, they should have enough filtration to produce an HVL of 3 mm.

PENETRATION WITH SCATTER

When a relatively large x-ray beam passes through a patient's body, as shown in Figure 10-7, some of the scattered radiation is added to the primary radiation penetrating the patient's body. This has the effect of increasing the apparent penetration or amount of radiation reaching the receptor. In fact, when most parts of the body are radiographed, there is actually much more scatter than primary radiation coming out of the patient's body in the direction of the receptor. This is undesirable because the scattered radiation reduces image contrast (this will be discussed in a later chapter). Also, the contribution from the scattered radiation must be taken into consideration when setting exposure techniques, especially when changing the field of view.

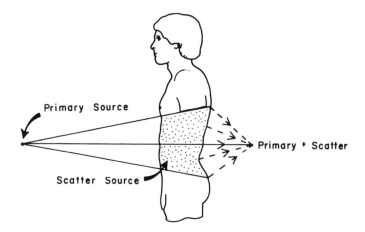

Figure 10-7 Scattered radiation adds to the primary radiation that penetrates an object.

The amount of scatter produced is determined by several factors. One of the most significant is the x-ray beam area, or field of view. Since the source of the scattered radiation is the volume of patient tissue within the area of the primary x-ray beam, the source size is proportional to the field of view. Another important factor is body section thickness, which also affects the size of the scattered radiation source. A third significant factor is KV. As the KV is increased over the diagnostic range, several changes occur. A greater proportion of the photons that interact with the body are involved in Compton interactions, and a greater proportion of the photons created in Compton interactions scatter in the forward direction. Perhaps the most significant factor is that the scattered radiation produced at higher KV values is more penetrating. A larger proportion of it leaves the body before being absorbed. When the scattered radiation is more penetrating, there is a larger effective source within the patient. At low KV values most of the scattered radiation created near the surface where the x-ray beam enters the body does not penetrate the body; at high KV values, this scattered radiation contributes more to the radiation passing through the body.

PENETRATION VALUES

We have seen that the amount of radiation that penetrates through a specific thickness of material is determined by the energy of the individual photons and the characteristics (density and atomic number) of the material. HVL values provide

Table 10-2 HVL Values for Certain Materials

Material	HVL (mm)		
	30 keV	60 keV	120 keV
Tissue	20.0	35.0	45.0
Aluminum	2.3	9.3	16.6
Lead	0.02	0.13	0.15

useful information about the penetration of a specific amount of radiation in a specific material. When an HVL value is known for a certain material, the penetration through other thicknesses can be easily determined. Table 10-2 gives HVL values for several materials related to diagnostic imaging.

Study Activities

Explain how HVL expresses the penetrating ability of an x-ray beam.

Determine the thickness of 1 HVL if 10 cm of a material reduces the radiation to 25% penetration.

Explain how you as an x-ray machine operator can change the HVL of an x-ray beam.

Explain the reason for having a filter in an x-ray beam.

Describe how a filter changes the spectrum of an x-ray beam.

Explain how increasing the amount of filtration affects the HVL of an x-ray beam.

Determine the thickness of material in HVLs required to reduce an exposure to less than 5%.

Chapter 11

Image Formation and Contrast

Contrast is the most important characteristic of a radiographic image. As we have said before, if there is no contrast, there is no image. An object in the body is visible only when it has contrast with respect to the area around it.

The contrast that ultimately appears in the image is determined by many factors, as indicated in Figure 11-1. In addition to the penetration characteristics to be considered here, image contrast is significantly affected by scattered radiation (Chapter 12) and the contrast characteristics of the film (Chapter 15). The contrast of small objects within the body and anatomical detail are reduced by image blurring (Chapter 18). As the x-ray beam emerges from the patient's body, as shown in Figure 11-2, it contains an image in the form of a difference in exposure between the object and background areas. A significant characteristic of this invisible x-ray image is the amount of contrast it contains. Contrast is represented by the difference in x-ray exposure between points within the image; the amount of contrast produced in a specific examination is determined by both the physical characteristics of the body section and the penetrating characteristics of the x-ray beam.

In this chapter we explore the characteristics of both the objects within a body and the x-ray beam and show how optimum image contrast can be achieved.

CONTRAST TYPES

Several types of contrast are encountered at different stages during the x-ray image formation. The formation of a visible image involves the transformation of one type of contrast to another at two stages in the image-forming process, as shown in Figure 11-2.

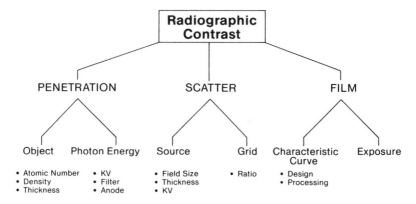

Figure 11-1 Factors that affect radiographic contrast.

Object Contrast

For an object to be visible in an x-ray image, it must have *physical contrast* in relationship to the tissue or other material in which it is embedded. This contrast can be due to a difference in *physical density* or *chemical composition* (atomic number).

When an object is physically different, it absorbs either more or less x-radiation than an equal thickness of surrounding tissue and casts a shadow in the x-ray beam. If the object absorbs less radiation than the surrounding tissue (ie, gas surrounded by tissue), it will cast a negative shadow that appears as a dark area in a radiograph. The third factor that affects contrast is object thickness in the direction of the x-ray beam. Object contrast is proportional to the product of object thickness and density. This quantity represents the mass of object material per unit area (cm^2) of the image. For example, a thick (large-diameter) blood vessel filled with diluted iodine contrast medium and a thin (small-diameter) blood vessel filled with undiluted medium will produce the same amount of contrast if the products of the blood vessels' diameters (or the objects *thicknesses*) and the concentration of the iodine within them (or the *density* of the iodine contrast medium) are the same.

The chemical composition of an object contributes to its contrast only if its effective atomic number is different from that of the surrounding tissue. Relatively little contrast is produced by the different chemical compositions found in soft tissues and body fluids because their effective atomic number values are close together. The contrast produced by a difference in atomic number is quite sensitive to photon energy as determined by KV_p. This is why changing the KV_p has an effect on contrast.

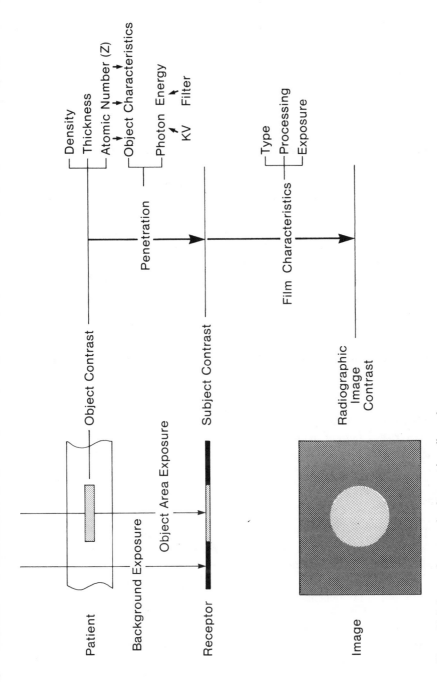

Figure 11-2 Stages of contrast development in radiography.

Most materials that produce high contrast with respect to soft tissue differ from soft tissue in both physical density and atomic number. The physical characteristics of most materials encountered in x-ray imaging are compared in Table 6-2 in Chapter 6.

Subject Contrast

The contrast in the invisible image emerging from the patient's body is traditionally referred to as *subject contrast*. Subject contrast is the difference in exposure between various points within the image area.

For an individual object, the significant contrast value is the difference in exposure between the object area and its surrounding background. This exposure difference is generally expressed as a percentage of the background exposure level. Contrast will be present if the exposure in the object area is either more or less than in the surrounding background.

Subject contrast is produced because x-ray penetration through an object in the body differs from penetration through the nearby background tissue. For objects that attenuate more of the radiation of an x-ray beam than the nearby tissue, contrast is inversely related to object penetration (ie, contrast increases as penetration is reduced). Maximum (100%) contrast is produced when no radiation penetrates the object. Metal objects (lead bullets, rods, etc) are good examples of objects producing maximum contrast. Contrast is reduced as x-ray penetration through the object increases. When object penetration approaches penetration through an equal thickness of surrounding tissue, contrast disappears.

The amount of subject contrast produced is determined by the physical contrast characteristics (atomic number, density, and thickness) of the object and the penetrating characteristics (photon energy spectrum) of the x-ray beam.

Image Contrast

The third type of contrast is the contrast that appears in the visible image. The contrast in a radiograph is in the form of differences in optical density between various points within the image, such as between an object area and the surrounding background. The amount of visible radiographic contrast produced in a specific procedure depends on the amount of x-ray beam exposure the receptor receives, which depends on the subject contrast, and the contrast transfer characteristics of the film, which will be discussed in Chapter 15.

The contrast in a visible fluoroscopic image is in the form of *brightness ratios* between various points within the image area. The amount of contrast in a fluoroscopic image depends on the amount of subject contrast entering the receptor system and the characteristics and adjustments of the imaging system. The contrast characteristics of a fluoroscopic system are discussed in Chapter 21.

EFFECT OF PHOTON ENERGY

Object penetration and the resulting contrast often depend on the photon energy spectrum. This in turn is determined by three factors: (1) x-ray tube anode material, (2) x-ray beam filtration, and (3) KV. Since most x-ray examinations are performed with tungsten anode tubes, the first factor cannot be used to adjust contrast. The exception is the use of molybdenum anode tubes in mammography. Most x-ray machines have essentially the same amount of filtration, which is a few millimeters of aluminum. Two exceptions are molybdenum filters used with molybdenum anode tubes in mammography and copper or brass filters sometimes used in chest radiography.

In most procedures KV_p is the only photon-energy controlling factor that can be changed by the operator to alter contrast. Radiographic examinations are performed with KV_p values ranging from a low of approximately 28 KV_p, in mammography, to a high of approximately 140 KV_p, in chest imaging. The selection of a KV_p for a specific imaging procedure is generally governed by the contrast requirement; other factors, such as patient exposure and x-ray tube heating will be considered later (Chapters 22 and 23).

Both photoelectric and Compton interactions contribute to the formation of image contrast. It was shown in Chapter 9 that the rate of Compton interactions is primarily determined by tissue density and depends very little on either tissue atomic number or photon energy. On the other hand, the rate of photoelectric interactions is very dependent on the atomic number of the material and the energy of the x-ray photons. This means that when contrast is produced by a difference in the atomic numbers of an object and the surrounding tissue, the amount of contrast is very dependent on photon energy (KV_p). If the contrast is produced by a difference in density (Compton interactions), it will be relatively independent of photon energy. Changing KV_p produces a significant change in contrast when the conditions are favorable for photoelectric interactions. In materials with relatively low atomic numbers (ie, soft tissue and body fluids), this change is limited to relatively low KV_p values. However, the contrast produced by materials with higher atomic numbers, such as calcium, iodine, and barium, will be affected by changing KV_p over a much wider range of KV_p values.

Soft Tissue Radiography

Two basic factors tend to limit the amount of contrast that can be produced between types of soft tissue and between soft tissue and fluid. One factor is the

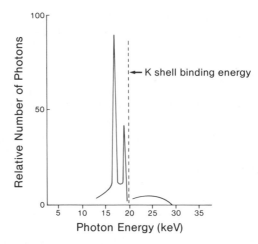

Figure 11-3 The x-ray spectrum used in mammography.

small difference in the physical characteristics (in terms of density and atomic numbers) of these materials, as shown in Table 6-1 in Chapter 6, and the second factor is the relatively low number of photoelectric interactions because of the low atomic numbers.

Mammography is a procedure that uses soft tissue contrast. The production of significant contrast requires the use of relatively low-energy photons. Mammography is typically performed with equipment that uses the characteristic radiation produced in a molybdenum anode x-ray tube and filtered by a molybdenum filter. The spectrum of this radiation is shown in Figure 11-3. The range of photon energies contained within this spectrum represents a reasonable compromise between contrast production and overall breast penetration (machine loading and patient exposure).

Calcium

Calcium produces significant contrast relative to soft tissue because it differs in both density and atomic number. Because of its higher atomic number, photoelectric interactions predominate over Compton interactions, up to a photon energy of approximately 85 keV. Above this energy the photoelectric interactions contribute little to image contrast.

Figure 11-4 shows the relationship between calcium penetration (contrast) and photon energy. In principle, the optimum photon energy range (KV_p) for imaging calcium depends, to some extent, on the thickness of the object. When imaging very small (thin) calcifications, as in mammography, a low photon energy must be

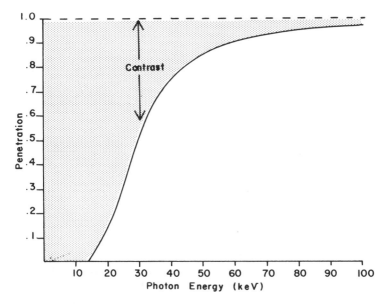

Figure 11-4 Relationship of calcium penetration and contrast to photon energy.

used or the contrast will be too low for visibility. When the objective is to see through a large calcified structure (bone), relatively high photon energies (KV_p) must be used to achieve adequate object penetration.

Iodine and Barium Contrast Media

The two chemical elements iodine and barium produce high contrast with respect to soft tissue because of their densities and atomic numbers. The significance of their atomic numbers ($Z = 53$ for iodine, $Z = 56$ for barium) is that they cause the K edge to be located at a very favorable energy relative to the typical x-ray energy spectrum. The K edge is 33 keV for iodine and 37 keV for barium. Maximum contrast is produced when the x-ray photon energy is slightly above the K edge energy of the material where penetration is low. This is illustrated for iodine in Figure 11-5. A similar relationship exists for barium but is shifted up to slightly higher photon energies.

Since the typical x-ray beam contains a rather broad spectrum of photon energies, all of the energies do not produce the same level of contrast. In practice, maximum contrast is achieved by adjusting the KV_p so that a major part of the spectrum falls just above the K edge energy. For iodine, this generally occurs when the KV_p is set in the range of 60 to 70 KV_p.

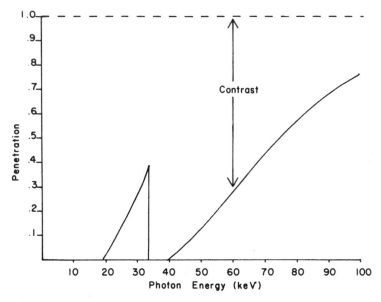

Figure 11-5 Relationship of iodine penetration and contrast to photon energy. The values shown are for a 1-mm thickness of iodine contrast medium.

AREA CONTRAST

We have considered a single object embedded in tissue. In this simple case an increase in contrast generally increases the visibility of the object. However, in most clinical applications one image contains many objects or anatomical structures. A problem arises when the different objects are located in different areas of the body and the thickness or density of the different body sections is significantly different. A chest image that contains lung and mediastinal areas is a good example; a simple representation of this is shown in Figure 11-6. Because of the large difference in tissue density between the lungs and the mediastinum, the contrast is significant between these two areas in the image. In this typical radiograph, the area of the mediastinum is very light (low film density), and the lung areas are much darker. Any objects within the mediastinum are imaged on a light background, and objects within the lung areas are imaged on dark backgrounds.

A characteristic of radiographic film is that its ability to display object contrast is reduced in areas that show up either very light (eg, the mediastinum) or relatively dark (eg, the lungs). If there is a relatively high level of contrast between *areas*

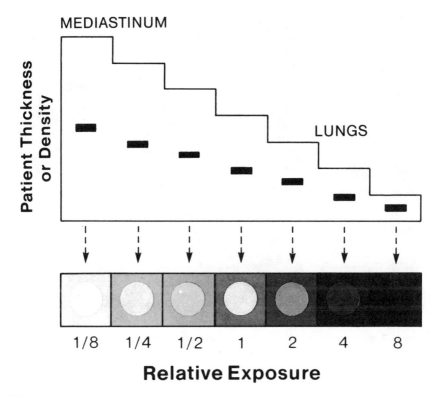

Figure 11-6 Physical conditions that produce area contrast.

within an image, then the *contrast of objects* within these areas can be reduced because of film limitations. Three actions can be taken to minimize the problem. One is to use a wide latitude film that reduces area contrast and improves visibility within the individual areas; this will be described in Chapter 15. A second approach is to place a compensating filter between the x-ray tube and the patient's body. The filter has areas with different thicknesses and is positioned so that its thickest part is over the thinnest, or least dense, part of the body. The overall effect is a reduction in area contrast within the image. The third action is to use a very penetrating x-ray beam produced by a high KV.

Figure 11-7 compares chest radiographs made at two KV_p values. The image on the left (A) was made at 60 KV_p. Although it has high contrast between the mediastinum and lung areas, visibility of structures within these areas is diminished. The high-KV radiograph (B), which has less area contrast, has increased object contrast, especially within the lung areas.

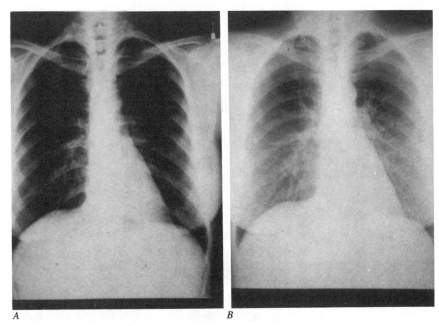

A   B

Figure 11-7 Radiographs illustrating a difference in area contrast produced by changing 60 KV$_p$ (A) to 140 KV$_p$ (B).

Study Activities

Name the three major factors that affect radiographic contrast.

Name the characteristics of an object in the body that determine its physical contrast.

Identify the primary physical characteristic that produces contrast between air in the lungs and soft tissue.

Explain why there is very little image contrast produced between fluid (water) and muscle.

Identify the physical characteristics that produce contrast between barium and soft tissue.

Explain how changing photon energy affects contrast between calcium and soft tissue.

Identify the KV$_p$ range that produces good contrast between iodine and soft tissue.

Explain why a high area contrast is often undesirable in a radiograph.

Explain the general relationship between area contrast and KV$_p$.

Chapter 12

Scattered Radiation and Contrast

When an x-ray beam enters a patient's body, a large portion of the photons engage in Compton interactions and produce scattered radiation. Some of this scattered radiation leaves the body in the same general direction as the primary beam and exposes the image receptor, reducing image contrast. The degree of loss of contrast depends on the scatter content of the radiation emerging from the patient's body. In most radiographic and fluoroscopic procedures the major portion of the x-ray beam leaving the patient's body is scattered radiation. This in turn significantly reduces contrast.

Subject contrast was previously defined as the difference in exposure delivered to the object area on a film, expressed as a percentage of the exposure delivered to the surrounding background. Maximum contrast (ie, 100%) is obtained when the object area receives no exposure with respect to the background. A previous chapter discussed the reduction of subject contrast because of x-ray penetration through the object being imaged. This chapter describes the further reduction of contrast by scattered radiation.

CONTRAST REDUCTION

The basic concept of *contrast reduction* by scattered radiation is illustrated in Figure 12-1. For simplicity's sake, it is assumed that the object is not penetrated and so, if it were not for scattered radiation, there would be 100% subject contrast. The object is assumed to be embedded in a larger mass of material, such as the human body, which is producing the scattered radiation. The exposure of the background area of the receptor has been produced by radiation that penetrates the body near to the object *plus* scattered radiation. The exposure to the object area has been produced by scattered radiation alone.

When an x-ray beam leaves a patient's body, it contains some primary radiation that has penetrated the body and some scattered radiation. For most body parts

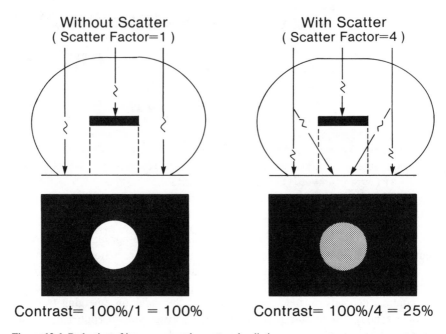

Figure 12-1 Reduction of image contrast by scattered radiation.

there is more scattered radiation than primary. The amount of scattered radiation coming out of a patient's body can be expressed by the *scatter factor*, S. The value of the scatter factor is the ratio of the total radiation (primary plus scatter) to the amount of primary radiation. If there were no scatter, the value of S would be 1. The value increases as the amount of scattered radiation increases. The amount of scatter, and value of the scatter factor, increases with the size and thickness of the body section being imaged. The limbs produce relatively little scatter, giving scatter factor values generally in the range of 1.5 to 3. An abdomen when imaged with a large field of view can produce scatter factors as high as 5 or 6.

The amount of contrast in an image is related to the amount of scattered radiation by

$$\text{Contrast with scatter} = \frac{\text{Contrast without scatter}}{\text{Scatter factor}}$$

$$Cs(\%) = 100/S$$

This relationship shows that as the proportion of scattered radiation in the x-ray beam increases, contrast proportionally decreases. For example, if the scatter factor has a value of 4, the contrast between the object and background areas will be reduced to 25%. In other words, the object area exposure is 75% of the exposure reaching the surrounding background. The background area exposure is, therefore, composed of one unit of primary and three units of scattered radiation. The object area receives only the three units of scattered radiation. This yields an object area exposure of 75% of background and a contrast of 25%.

With respect to image contrast, the scatter factor, S, is also termed the *contrast reduction factor*. For example, if the scatter factor has a value of 2, the resulting contrast will be 50%. This is a reduction of 100% contrast by a factor of 2. A scatter factor value of 5 reduces contrast by a factor of 5, or down to 20%. Figure 12-2 shows the general relationship between contrast and scatter factor. The value of the scatter factor is primarily a function of patient thickness, field size, and KV_p. In examinations of relatively thick body sections, contrast reduction factors of 5 or 6 are common.

We developed our discussion of contrast reduction using an unpenetrated object that would produce 100% contrast in the absence of scatter. But most objects within the body are penetrated to some extent. Therefore, contrast is reduced by both object penetration and scattered radiation. For example, if an object is 60% penetrated (40% contrast), and the scatter factor has a value of 4, the final contrast will be only 10%.

Since scattered radiation robs an x-ray image of most of its contrast, specific actions must be taken to regain some of the lost contrast. Several methods can be used to reduce the effect of scattered radiation, but none is capable of restoring the full image contrast. The use of each scatter reduction method usually involves compromises, as we will see below.

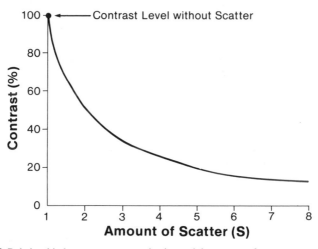

Figure 12-2 Relationship between contrast reduction and the amount of scatter.

COLLIMATION

The amount of scattered radiation is generally proportional to the total mass of tissue contained within the primary x-ray beam. This is in turn determined by the thickness of the patient and the area or field size being exposed. Increasing the field size increases the total amount of scattered radiation and the value of the scatter/contrast-reduction factors. Therefore, one method of reducing scattered radiation and increasing contrast is to reduce the field size with x-ray beam collimators, cones, or other beam-limiting devices, as illustrated in Figure 12-3. This method is limited by the necessity to cover a specific anatomical region. However, in most situations, contrast can be improved by reducing the field size to the smallest practical value.

AIR GAP

The quantity of scattered radiation reaching a receptor can be reduced by separating the patient's body and the receptor surface, as shown in Figure 12-4. This separation is known as an *air gap*. Scattered radiation leaving a patient's body is less directed than the primary x-ray beam. Therefore, scattered radiation spreads out of the primary beam area. The proportion of scattered radiation to primary radiation at the image receptor increases with air gap distance. Several factors must be considered when using this method of scatter reduction. Patient exposure is increased because of the inverse-square effect. The use of an air gap magnifies

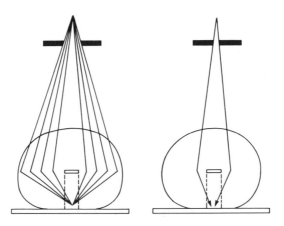

Figure 12-3 Contrast improvement by reducing x-ray beam size.

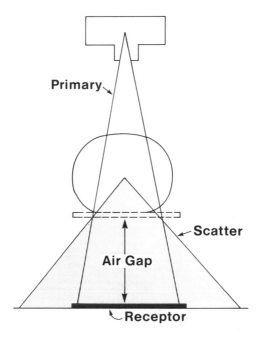

Figure 12-4 Contrast improvement by using an air gap.

the image size, so that a larger receptor size is required to obtain the same patient area coverage. If the air gap is obtained by increasing the tube-to-receptor distance, the x-ray equipment must be operated at a higher output to obtain adequate receptor exposure.

Also, increasing the separation distance between the patient and the receptor increases focal spot blurring. It is usually necessary to use relatively small focal spots with an air-gap technique.

Study Activities

Explain how scattered radiation reduces the contrast of an object within the body.

Identify the major factor that determines the amount of scattered radiation produced in a patient's body.

Explain how you can reduce the amount of scattered radiation coming from a patient's body.

Explain how the air gap technique reduces the intensity of scattered radiation to the receptor.

Chapter 13

Grids

In the previous chapter we learned that the scattered radiation produced in a patient's body can add exposure to the film and reduce the image contrast.

Most radiographic procedures use a *grid* to stop some of the scattered radiation before it reaches the receptor. The grid is in the form of a thin, flat sheet, which is placed between the patient's body and the receptor. The grid is designed so that it lets most of the primary x-ray beam (good radiation) pass through, but it blocks most of the scattered (bad) radiation.

In this chapter we will first learn how the grid works. Then we will see why there are different types of grids and how they should be used.

GRID FUNCTION

The function of a grid is to absorb as much of the scattered radiation as possible and as little of the primary x-ray beam as possible. Therefore, the grid must be constructed so that the primary radiation can pass through but not the scattered radiation. If we look at a cross-section of a grid, as shown in Figure 13-1, we see how this can happen.

The grid is made up of alternating strips of two different materials. One material, usually lead, is a very good x-ray absorber. The other material, usually aluminum or fiber, is not a good absorber, and the radiation can pass through it. In the illustration, the lead strips are black, and the aluminum or fiber is represented by a gap between the lead strips. In the illustration the grid looks much thicker than it actually is. This is so that we can see the details of its construction.

Grid Ratio

One of the important characteristics of a grid is its ratio. The *grid ratio* is the ratio of two dimensions found within the grid. One is the thickness of the grid, t,

127

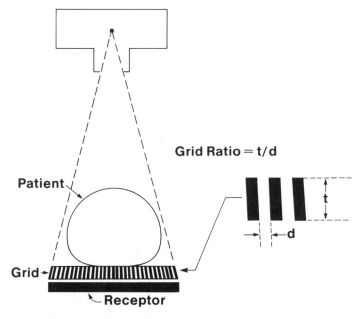

Figure 13-1 Contrast improvement by using a grid.

and the other is the width of the gap, d. A grid is manufactured with a specific ratio value, usually in the range of 5 to 1 up to 16 to 1. You can determine the ratio of the grid you are using by looking at the label.

GRID ABSORPTION

Primary Beam Penetration

The ideal grid would be one that allows all of the primary x-ray beam to pass through without being absorbed. Although the real grids we use are not perfect in this respect, they do allow most of the primary beam to pass through without being absorbed. This is because the direction of the primary beam is lined up with the gaps, as shown in Figure 13-2. Unfortunately, some of the radiation will hit the lead strips and be absorbed, but this will be a relatively small percentage if the grid is properly aligned with the x-ray beam. Most grids will let between 60 to 70% of the primary radiation pass through.

When using a grid, the percentage of primary radiation that penetrates should be as high as possible. Therefore, it is necessary that the grid be properly positioned and aligned with respect to the x-ray beam. We will come back to this point later.

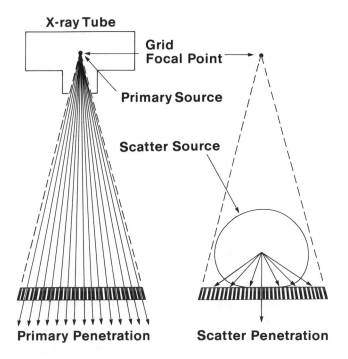

X-ray Tube

Grid Focal Point

Primary Source

Scatter Source

Primary Penetration

Scatter Penetration

Figure 13-2 Selective absorption of scattered radiation by a grid.

Scatter Absorption

The primary function of a grid is to absorb the scattered radiation coming from the patient's body. Because the body (the source of scattered radiation) is located closer to the grid than to the x-ray tube (the source of primary radiation), most of the scattered radiation will not be lined up with the gaps. Most of it will hit the lead strips at an angle and be absorbed, as shown in Figure 13-2. That's good. A small amount of the scattered radiation will be going in a direction that allows it to pass through the gaps and reach the receptor. This is bad. However, for most grids this is less than 10%.

Bucky Factor

Technique factors must be adjusted to compensate for the radiation absorbed by the grid. We must use more exposure (KV or MAS) when a grid is used. The characteristic of the grid that tells us how much to increase the exposure is the

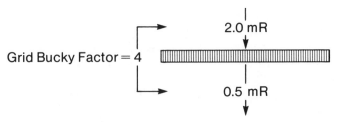

Figure 13-3 The concept of the Bucky factor.

Bucky factor. The concept of the Bucky factor is illustrated in Figure 13-3. The value of the Bucky factor is the ratio of the amount of radiation, both primary and scatter, going into the grid compared to the amount that comes out on the other side.

The value of the Bucky factor is determined by two things: the grid ratio and the amount of scatter coming from the patient's body. High-ratio grids that absorb more radiation than low-ratio grids will have higher Bucky factor values. However, a grid that has a specific grid ratio will have a range of Bucky factor values, depending on the amount of scattered radiation from the patient's body.

Therefore, when radiographing large body sections that produce much scatter, the Bucky factor value will be higher than when the same grid is used on a smaller body part or with a smaller field of view. The value of the Bucky factor is also affected by KV. A grid's Bucky factor increases with KV because more scatter comes from the patient's body.

It is not necessary to know precise Bucky factor values in order to select technique factors for a specific patient. The important thing is to understand how the use of a grid affects technique factors in general and why we must use more exposure with a high ratio grid that has a higher Bucky factor value.

Typical Bucky factor values are given below.

Grid Ratio	Bucky Factor
(No grid)	1
5:1	2–3
8:1	3–4
12:1	4–5
16:1	4–6

GRID FOCUS

We have already seen that the primary x-ray beam can pass through the grid because it is lined up with the gaps between the lead strips. This works best when the strips and gaps line up with a single point in space—the point where the radiation is coming from. A grid built this way is called a *focused grid* because the gaps focus on a single point in space, as shown in Figure 13-2.

In order to get the most primary radiation through the grid, it is necessary that the grid be properly aligned with the x-ray beam so that the *grid focal point* is at the same location as the x-ray tube focal spot. Therefore, we need to know where the grid focal point is located. This information is on the grid label.

The label will indicate the distance from the grid surface to the focal point. The grid will also be marked to indicate which side of the grid should be turned toward the x-ray tube.

If a grid is not positioned so that its focal point is close to the x-ray tube focal spot, *grid cut-off* will occur. In this condition, the primary x-ray beam cannot pass through the grid in certain areas. We will come back to this problem when we discuss grid alignment later.

It is not necessary for the focal point of the grid to be precisely at the x-ray tube focal spot. Instead, the tube focal spot may be located anywhere within an area around the grid focal point, to prevent grid cut-off. The size of this area depends on the ratio of the grid. A high-ratio grid has a smaller area than a low-ratio grid. This means the positioning and alignment of a high-ratio grid must be more precise than for a low-ratio grid.

Each focused grid has a range of distances in which the x-ray tube focal spot should be located. A grid's range is indicated on its label. A low-ratio grid has a much larger range than a high-ratio grid. The two characteristics, focal distance and range, should not be confused. Focal *distance* indicates the distance from the grid to the grid's focal point. The *range* indicates how close the grid focal point must be located to the x-ray tube focal spot for proper operation.

Parallel Strip Grids

Some grids are designed with the strips running parallel to one another rather than being focused on a point. A parallel grid has no specific focal point and can be used over a relatively wide range of distances from the x-ray tube.

The use of parallel grids is limited by cut-off near the edges of the field. However, they can be used satisfactorily for relatively small fields of view or when there is a long distance to the x-ray tube.

CROSSED GRIDS

Some grids are designed with two sets of strips that cross each other. This is compared to the linear grid we have discussed up to this point.

The advantage of a crossed grid is that it absorbs more scattered radiation than a linear grid and produces better cleanup.

The disadvantage in using a crossed grid is that it requires a higher exposure, and the x-ray beam cannot be angled without producing grid cut-off.

GRID ALIGNMENT AND CUT-OFF

A grid will work properly only when it is aligned with the x-ray beam so that it does not produce unnecessary absorption or cut-off. A grid is properly aligned when the x-ray tube focal spot is within the grid's focal range. There are several types of misalignment that can occur, as shown in Figure 13-4.

Distance

If the distance between the grid and the x-ray tube is not within the range, the primary x-ray beam will not be properly aligned near the edges of the grid, and cut-off will occur there. This can result in decreased exposure along the edges of the radiograph. Distance misalignment does not usually affect the exposure in the center of the radiograph.

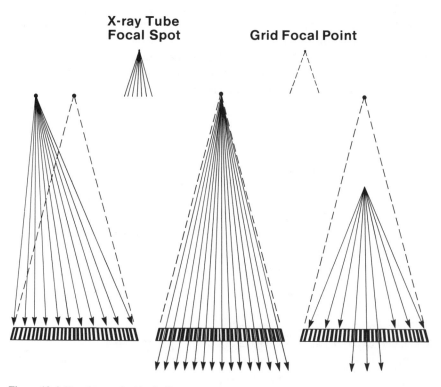

Figure 13-4 Two forms of grid misalignment.

Off-Center

If the x-ray tube focal spot is not properly centered with the grid, cut-off will occur over the entire grid area.

Upside-Down

If a grid is positioned so that its focal point (tube side) is on the side away from the x-ray tube, very serious cut-off will occur. Radiographers should be very careful when changing grids to make sure the correct side of the grid is facing the x-ray tube.

GRID LINES

In many cases the lead strips in a grid will produce visible lines in the radiograph. This is undesirable.

One method used in many radiographic systems to eliminate grid lines in the image is to blur them by moving the grid during the exposure. The mechanism for accomplishing this was first introduced by Dr. Hollis Potter, and a moving grid system is often referred to as a *Potter-Bucky diaphragm*. In a Potter-Bucky system, the grid moves at right angles to the grid lines. The speed at which the grid moves determines the shortest exposure time that will not produce grid lines.

GRID SELECTION

The various characteristics of a grid that we have discussed must be considered when selecting a grid for a specific clinical application.

Study Activities

Define grid ratio.

Explain why most of the primary x-ray beam can penetrate a grid.

Explain why most of the scattered radiation is absorbed by a grid.

Explain the general relationship between grid penetration and grid ratio.

Define Bucky factor.

Describe the type of cutoff produced if the x-ray tube focal-spot-to-receptor distance is much shorter than the focal distance of the grid.

Identify the general conditions in which the use of a low-ratio grid is appropriate.

Identify the general conditions in which the use of a high-ratio grid is appropriate.

Chapter 14

The Photographic Process and Film Sensitivity

Most radiographs are recorded on film. The image is created by the same basic photographic process used to make a black and white snapshot with a camera. The difference in a radiograph is actually in the negative, which is in the form of a transparency. In photography we view a positive print that has been made from the negative. To make an image on film requires two distinct steps. The first step is to expose the film to radiation, such as light, and the second step is to process the film. When the film is exposed, an invisible or *latent image* is created in the film. Processing causes the latent image to become visible.

All radiographic films do not have the same characteristics. Therefore, we must know something about a film when we are selecting it for a specific clinical procedure. There are two very important characteristics for each film. These are (1) sensitivity and (2) contrast. In this chapter we will look at the basic process of creating an image and the factors that affect a film's sensitivity. The contrast characteristics of film will be discussed in the next chapter.

FILM FUNCTIONS

It is helpful if we recognize that film actually performs several different functions for us in radiography.

Image Recording

In principle, film is an image converter. It converts radiation, typically light, into various shades of gray. An important characteristic of film is that it records, or retains, an image. An exposure of a fraction of a second can create a permanent image. The amount of exposure required to produce an image depends on the sensitivity (speed) of the film being used. Some films are more sensitive than others because of their design or the way they are processed. The sensitivity of radiographic film is generally selected to provide a compromise between two very important factors: patient exposure and image quality. A highly sensitive film reduces patient exposure but decreases image quality.

Image Display

Most medical images are recorded as transparencies. In this form they can be easily viewed by illumination from a view box. The overall appearance and quality of a radiographic image depends on a combination of factors, including the characteristics of the particular film used, the way in which it was exposed, and the processing conditions. When a radiograph emerges from the film processor, the image is permanent and cannot be changed. So, it is important that everything associated with the production of the image be adjusted to produce the best image quality.

Image Storage

Film has been the traditional medium for medical image storage. If a film is properly processed, it will have a lifetime of many years and will, in most cases, outlast its clinical usefulness. The major disadvantages of storing images on film are its bulk and the difficulties of filing it. Most clinical facilities must devote considerable space to film storage. Retrieving films from storage generally requires manual search and transportation of the films to a viewing area.

Because film performs so many of the functions that make up the radiographic examination, it will continue to be an important element in the medical imaging process. Because of its limitations, however, it will gradually be replaced by digital electronic imaging media in many clinical applications.

OPTICAL DENSITY

When you look at a radiograph on a view box, you see areas with different degrees of brightness. The variation in brightness is determined by how much light comes through the film. In a dark area very little light passes through the film. The characteristic of a film that determines the amount of light coming through a specific point is the *optical density*. Density is the term we use to describe the darkness or opaqueness of film. Each point in a radiograph has a specific density value. Light (bright) areas have low density values, whereas dark areas have high density values.

The optical density value is determined by the fraction or percent of the view box light that passes through the film. Let's use Figure 14-1 to look at this relationship.

A clear piece of film that allows 100% of the light to penetrate has a density value of 0. Radiographic film is never completely clear. The minimum film density is usually in the range of 0.1 to 0.2 density units. This is designated the *base plus fog density* and is the density of the film base and any inherent fog. Fog is any density in the film that is not produced by the exposure that creates the image. The various sources of fog are described later. There is always a small amount of inherent fog even when the film has received no exposure.

Each unit of density decreases light penetration by a factor of 10. A film area with a density value of 1 allows 10% of the light to penetrate and generally appears as a medium gray when placed on a conventional view box. A film area with a density value of 2 allows 10% of 10% (1%) light penetration and appears as a relatively dark area when viewed in the usual manner. With normal view box illumination, it is possible to see through areas of film with density values of up to approximately 2 units.

A density value of 3 corresponds to a light penetration of 0.1% (10% of 10% of 10%). A film area with a density value of 3 appears essentially opaque when illuminated with a conventional view box. It is possible, however, to see through such a film by using a bright, "hot" light. Radiographic film generally has a maximum density value of approximately 3 density units. This is designated the D_{max} of the film. The maximum density that can be produced within a specific film depends on the characteristics of the film and processing conditions.

Measurement

The density of film is measured with a *densitometer*. A light source passes a small beam of light through the film area to be measured. On the other side of the film, a light sensor (photocell) converts the penetrated light into an electrical signal. A special circuit processes the signal and displays the results in density units.

The primary use of densitometers in a clinical facility is to monitor the performance of film processors.

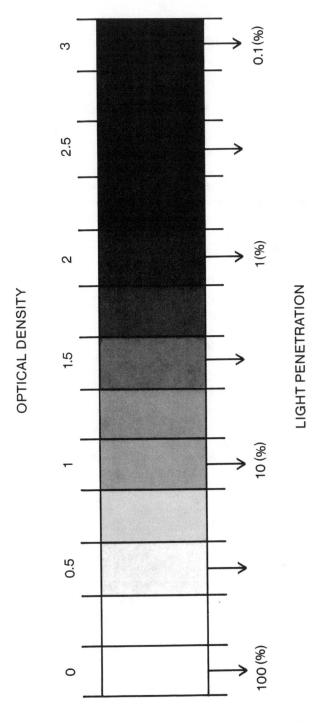

Figure 14-1 Relationship between light penetration and film density.

Fog

All of the density in a radiograph is not always "good" density that contributes to the image. There is often some "bad" density, which we know as *fog*. The presence of fog reduces the contrast and quality of the image.

There are several things that can cause fog:

1. *Light exposure:* Accidental exposure to light is a common cause of fog. This can come from cassettes that do not close tightly, light leaks into darkrooms, incorrect safelights in the darkroom, or opening a film container in a lighted area.

2. *X-ray exposure:* Film is sensitive to x-radiation as well as light. Therefore, we should not keep film in an x-ray room unless it is properly shielded against the radiation.

3. *Time:* Unprocessed film will develop some fog with age. Radiographic film should not be stored for long periods of time before use.

4. *Heat:* Film will age (develop fog) faster as the temperature is increased. Storage areas should be as cool as possible.

5. *Chemical action:* Unwanted density (fog) can be caused by common chemical action. The most common form is overdeveloping, which causes some of the unexposed film grains to turn dark and produce fog density. There are several things that can cause a film to be overdeveloped and fogged, as we will see in Chapter 16.

FILM STRUCTURE

Conventional film is layered, as illustrated in Figure 14-2. The active component is an *emulsion* layer coated onto a *base* material. Most film used in radiography has an emulsion layer on each side of the base so that it can be used with two intensifying screens simultaneously. Films used in cameras and in selected radiographic procedures, such as mammography, have one emulsion layer and are called single-emulsion films.

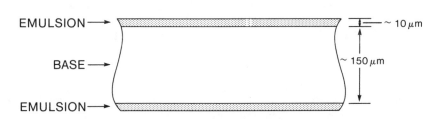

EMULSION ⟶   ~ 10 μm

BASE ⟶   ~ 150 μm

EMULSION ⟶

Figure 14-2 Cross-section of typical radiographic film.

Base

The base of a typical radiographic film is made of a clear polyester material that is 150 to 200 μm thick. It provides the physical support for the other film components and does not participate in the image-forming process. In some films the base contains a light blue dye to give the image a more pleasing appearance when illuminated on a view box.

Emulsion

The emulsion is the active component in which the image is formed and consists of many small silver halide crystals, which are called *grains*, suspended in gelatin. The gelatin supports, separates, and protects the crystals. The typical emulsion is 5 to 10 μm thick.

Silver halides are chemical compounds that contain the element silver. Several different silver halides have photographic properties, but the one typically used in medical imaging films is silver bromide. The silver bromide is in the form of crystals, each containing on the order of 10^9 atoms.

Silver halide grains are irregularly shaped, like pebbles or grains of sand. Several grain shapes are used in film emulsions. For many years radiographic film used a cubic grain configuration, in which the three dimensions of the cube-shaped grains were approximately equal. More recently, tabular-shaped grains were developed. The tabular grain is relatively thin in one dimension; its length and width are much larger than its thickness, giving it a relatively large surface area. The primary advantage of tabular grain film in comparison to cubic grain film is that dyes can be used more effectively in the emulsion to increase sensitivity and reduce crossover exposure. This will be described more fully later.

THE PHOTOGRAPHIC PROCESS

The production of film density and the formation of a visible image is a two-step process. The first step in this photographic process is the exposure of the film to light, which forms an invisible *latent image*. Latent means the image is actually present in the emulsion but not in visible form. The second step is the chemical process that converts the latent image into a visible image with a range of densities (shades of gray).

Film density is produced by converting silver ions into metallic silver, which causes each processed grain to become black. The process is rather complicated and is illustrated by the sequence of events shown in Figure 14-3.

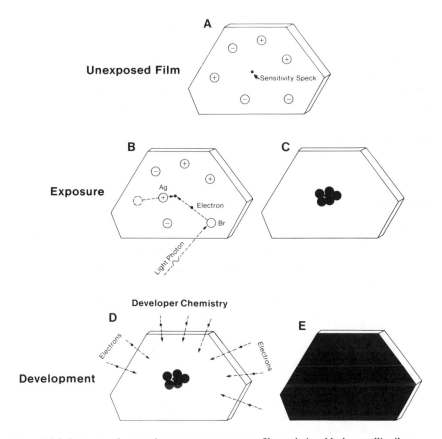

Figure 14-3 Sequence of events that convert a transparent film grain into black, metallic silver.

Each film grain contains a large number of both silver (Ag) and bromide (Br) ions. The silver ions have a one-electron deficit, which gives them a positive charge. On the other hand, the bromide ions have a negative charge because they contain an extra electron. Each grain has a structural "defect" known as a "sensitive speck." A film grain in this condition is relatively transparent.

Latent Image Formation

The first step in the formation of the latent image is the absorption of light photons by the bromide ions, which frees the extra electron. The electron moves to the sensitivity speck, causing it to become negatively charged. The speck in turn attracts one of the positively charged silver ions. When the silver ion reaches the speck, its positive charge is neutralized by the electron. If this process is repeated several times within an individual grain, the cluster of metallic silver at the sensitive speck will become a permanent arrangement. This action converts the silver ion into an atom of black metallic silver. The number of grains in the emulsion that reach this status depends on the overall exposure to the film. The grains that received sufficient exposure to form a permanent change are not visually different at this stage from the unexposed grains but are more sensitive to the action of the developer chemistry. The distribution of these activated but "invisible" grains throughout the emulsion creates the latent image.

Development

The invisible latent image is converted into a visible image by the chemical process of *development*. The developer solution supplies electrons that migrate into the sensitized grains and convert the other silver ions into black metallic silver. This causes the grains to become visible black specks in the emulsion.

The gray or black density that we see in a film is just a collection of many black grains of silver. When the concentration of the grains is increased, the density is increased.

SENSITIVITY

One of the most important characteristics of film is its *sensitivity*, often referred to as film speed. The sensitivity of a particular film determines the amount of exposure required to produce an image. A film with a high sensitivity requires less exposure than a film with a lower sensitivity.

The sensitivities of films are generally compared by the amount of exposure required to produce an optical density of 1 unit above the base plus fog density. The sensitivity of radiographic film is generally not described with numerical values but rather with a variety of generic terms such as "half speed," "medium speed," and "high speed." Radiographic films are usually considered in terms of their relative sensitivities (speeds) rather than their absolute sensitivity values. Although it is possible to choose films with different sensitivities, the choice is limited to a range of about four to one by most manufacturers.

Figure 14-4 compares two films with different sensitivities. Notice that a specific exposure produces a higher density in the high-sensitivity film; therefore, the production of a specific density value (ie, 1 density unit) requires less exposure.

High-sensitivity films are chosen when the reduction of patient exposure and the heat loading of the x-ray equipment are important considerations.

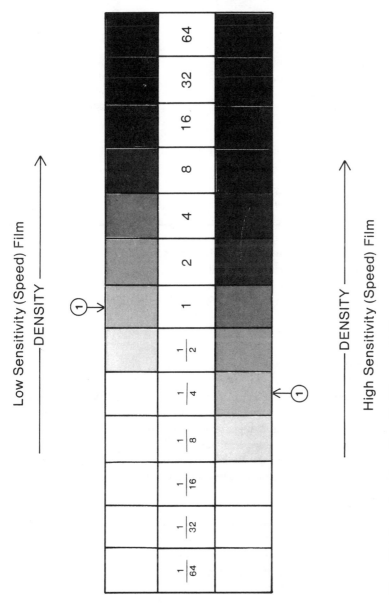

Figure 14-4 Comparison of two films with different sensitivities.

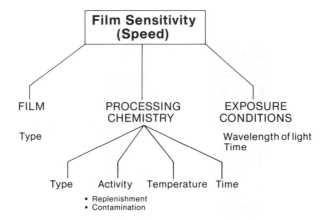

Figure 14-5 Factors that affect film sensitivity.

Low-sensitivity films are used to reduce image noise. The relationship of film sensitivity to image noise is considered in Chapter 19.

The sensitivity of film is determined by a number of factors, as shown in Figure 14-5, including its design, the exposure conditions, and how it is processed.

Composition

The basic sensitivity characteristic of a film is determined by the composition of the emulsion. The size and shape of the silver halide grains have some effect on film sensitivity. Increasing grain size generally increases sensitivity. Tabular-shaped grains generally produce a higher sensitivity than cubic grains. Although grain size may vary among the various types of radiographic film, most of the difference in sensitivity is produced by adding chemical sensitizers or dyes to the emulsion.

Processing

The effective sensitivity of film depends on several factors associated with the development:

- the type of developer
- developer concentration

- developer replenishment rates
- developer contamination
- development time
- development temperature

In most medical imaging applications, the objective is not to use these factors to vary film sensitivity but rather to control them to maintain a constant and predictable film sensitivity.

In Chapter 16 we will go into much more detail of the development process.

LIGHT COLOR

Film is not equally sensitive to all colors of light. For example, a specific film might be more sensitive to the light in the blue than in the green part of the spectrum. *Spectral sensitivity* is a characteristic of film that must be taken into account in selecting film for use with specific intensifying screens and cameras. In general, the film should be most sensitive to the light that is emitted by the intensifying screens, intensifier tubes, cathode ray tubes (CRTs), or lasers.

Blue Sensitivity

A basic silver bromide emulsion has its maximum sensitivity to light in the ultraviolet and blue regions of the spectrum. For many years most intensifying screens contained calcium tungstate, which emits blue light and is a good match for blue-sensitive film. Although calcium tungstate is no longer widely used as a screen material, several contemporary screen materials emit blue light and must be used with a blue-sensitive film. Discussion of intensifying screens continues in Chapter 17.

Green Sensitivity

Several image light sources, including image intensifier tubes, CRTs, and some intensifying screens, emit most of their light in the green portion of the spectrum. This will be discussed in more detail later. Film used with these devices must therefore be sensitive to green light. Silver bromide can be made sensitive to green light by adding special sensitizing dyes to the emulsion.

Users must be careful not to use the wrong type of film with intensifying screens. If a blue-sensitive film is used with a green-emitting intensifying screen, the combination will have a drastically reduced sensitivity.

Red Sensitivity

Many lasers produce red light. Devices that transfer images to film by means of a laser beam must therefore be supplied with a film that is sensitive to red light.

Safelighting

Darkrooms in which film is loaded into cassettes and transferred to processors are usually illuminated with a safelight. A safelight emits a color of light that the eye can see but that will not expose film. Although film has a relatively low sensitivity to the light emitted by safelights, film fog can be produced with safelight illumination under certain conditions. The safelight should provide sufficient illumination for darkroom operations but not produce significant exposure to the film being handled. This can usually be accomplished if certain factors are controlled. These include safelight color, brightness, location, and duration of film exposure.

The color of the safelight is controlled by the type of light bulb or the color of the filter placed over the bulb. The filter must be selected in relationship to the spectral sensitivity of the film being used. The different filter colors are designated by type numbers that are standard in the industry. An amber-brown safelight provides a relatively high level of working illumination and adequate protection for blue-sensitive film; type 6B filters are used for this application. However, this type of safelight produces some light that falls within the sensitive range of green-sensitive film.

A red safelight is required when working with green-sensitive films. Type GBX filters are used for this purpose.

Selecting the appropriate safelight filter does not absolutely protect film, because film has some sensitivity to the light emitted by most safelights. Therefore, the brightness of the safelight (determined by bulb size) and the distance between the light and film work surfaces must be selected so as to minimize film exposure.

Since exposure is an accumulative effect, handling the film as short a time as possible minimizes exposure. The potential for safelight exposure can be evaluated in a darkroom by placing a piece of film on the work surface, covering most of its area with an opaque object, and then moving the object in successive steps to expose more of the film surface. The time intervals should be selected to produce exposures ranging from a few seconds to several minutes. After the film is processed, the effect of the safelight exposure can be observed. Film is most sensitive to safelight fogging *after* the latent image is produced but *before* it is processed.

EXPOSURE TIME

In radiography it is usually possible to deliver a given exposure to film by using many combinations of radiation intensity (exposure rate) and exposure time. Since radiation exposure rate is proportional to x-ray tube MA, this is equivalent to saying that a given exposure (MAS) can be produced with many combinations of MA and time (S). This is known as the *law of reciprocity*. In effect, it means that it is possible to swap radiation intensity (MA) for exposure time and produce the same film exposure. When a film is directly exposed to x-radiation, the reciprocity law holds true. That is, 100 mAs will produce the same film density whether the film is exposed at 1,000 mA and 0.1 seconds or 10 mA and 10 seconds. However, when a film is exposed by light, such as from intensifying screens or image intensifier tubes used in fluoroscopy, the reciprocity law does not hold. With light exposure, as opposed to direct x-ray interactions, a single silver halide grain must absorb more than one photon before it can be developed and can contribute to image density. This causes the sensitivity of the film to be somewhat dependent on the intensity of the exposing light. This loss of sensitivity varies to some extent from one type of x-ray film to another. The clinical significance is that MAS values that give the correct density with some exposure times might not do so with other exposure times even when the MAS and exposure is the same. This is known as reciprocity law *failure*.

Study Activities

Define latent image.

Identify the step that converts a latent image into a visible image.

Describe the general relationship between radiation exposure and the resulting optical density of film.

Describe the general relationship between optical density and the percentage of light that penetrates the film.

Explain the basic function performed by the base of a film.

Explain the basic function performed by the film emulsion.

Briefly describe the steps by which a film grain is converted into black metallic silver.

Identify the three major factors that affect the sensitivity (speed) of film.

Explain the expected result of exposing a blue-sensitive film to an intensifying screen that produces green light.

Explain why a safe light does not generally expose x-ray film.

Explain reciprocity law failure as it applies to x-ray film.

Identify four potential sources of film fog.

Chapter 15

Film Contrast Characteristics

Contrast is perhaps the most important characteristic of a radiographic image. As we have said before, if you don't have contrast, you don't have an image. Contrast, or different density values, is what makes an image. The visibility of objects within a patient's body is determined by the amount of contrast produced in the radiograph. We have already discovered two factors that affect contrast. One is the amount of radiation penetrating the object. The other is the effects of scattered radiation. We now come to the third major factor that affects radiographic contrast: the film. One of the problems we face in radiography is that under certain conditions film does not produce enough contrast for good visibility.

RELATIONSHIP OF EXPOSURE AND CONTRAST

Film can be considered a contrast converter. One of its functions is to convert or transfer differences in exposure (subject contrast) into film contrast (differences in density), as shown in Figure 15-1. Exposure contrast is the difference in exposure between two areas of an image. The amount of film contrast produced by a certain exposure contrast can vary considerably. It depends on many factors, including the design characteristics of the film, processing conditions, and the amount of exposure delivered to the film.

Let's begin by thinking about the relationship between exposure contrast and film contrast. Figure 15-1 shows two areas of film that have received different exposures. The area on the left has received one half (50%) of the exposure that the area on the right has received. We can call this an exposure contrast of 50%. Let us assume that the film is now processed and that the density values for the two areas are measured and found to be 0.8 and 1.4 density units. The amount of film contrast between the two areas is the difference between these density values, or 0.6. The amount of film contrast produced by a 50% exposure contrast is known as

149

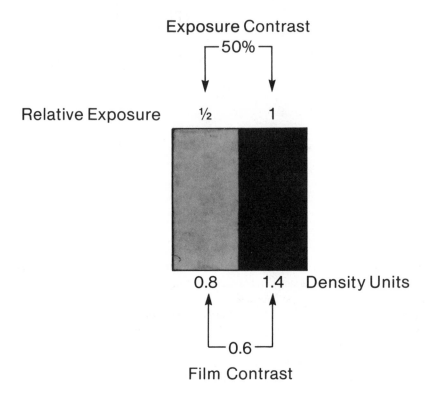

Figure 15-1 The general relationship between exposure contrast and film contrast.

the film's *contrast factor*. The value of the contrast factor tells us the amount of contrast that a film produces. The problem is that film does not always produce the same amount of contrast. There are many times when the contrast will be much less and not sufficient for the visibility of some objects in the body. Let's find out how this can happen.

We will now perform an experiment on a piece of radiographic film to find out how good it is at producing contrast. We will use a strip of film, as shown in Figure 15-2. The strip of film will be divided into small areas as shown. We will now expose each area to a different amount of radiation. Above the strip we have a scale showing the relative exposure from one area to another. An important thing about the exposure is that we have always changed it by the same ratio (percentage) as we have moved from one step to another. As we move from left to right, the exposure is always being doubled, or increased by a factor of two. If we look from the right to the left, the exposure is being decreased by a half, or 50%. If we now select any two adjacent steps, anywhere along the strip, the two steps will have received a 50% exposure contrast relative to each other.

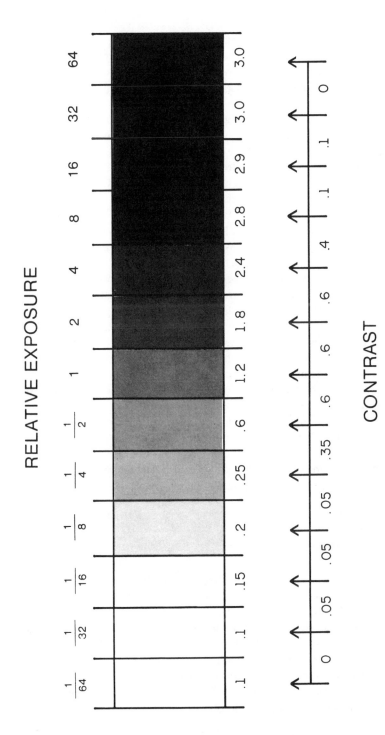

Figure 15-2 The variation in contrast with exposure.

Exposing a strip of film in this manner is done by using a device known as a *sensitometer*. The sensitometer is most frequently used to expose a strip of film for checking the performance of the film processor, as we will discuss later. But it is also useful for studying the contrast characteristics of the film, as we are doing here.

As a next step in our experiment, we will process the strip of film and then read the density values for each step. The density values are indicated immediately below each step. We notice that the density increases as the exposure increases from step to step. We see it ranging from a low of 0.1 on the left up to a value of 3.0 in the high-exposure area on the right. This should come as no surprise, because we have already learned that film density is increased as we increase the exposure to the film.

Let's now begin to look at the film contrast between each pair of steps (the bottom most scale in Figure 15-2). We can find the contrast factor value by taking the difference (subtracting) between the density values. Let's start on the left. We find that there is no contrast between the first two steps, because they have the same density value. What has occurred is that the exposure of this area of the film is too low to produce contrast. This demonstrates one of the problems with film. If it is underexposed, it does not produce adequate contrast. Let's now work up the scale and look at the contrast between other pairs of steps. We notice that as the exposure is increased, the contrast also increases, up to a maximum value of 0.6. This occurs near the middle of the exposure range. However, as we continue up the exposure scale and the film gets darker (more dense), we discover that the contrast begins to decline and eventually drops back to zero.

This simple experiment has revealed to us a very important characteristic of radiographic film. The amount of contrast the film produces depends on the amount of exposure to the film. For every type of film there is an exposure value that produces maximum contrast. If areas in a radiograph receive exposures that are either less than or higher than this optimum value, the amount of contrast produced by the film will be reduced.

LOSS OF CONTRAST

In radiography there are two general factors that cause film to be either under- or overexposed and not produce its maximum contrast. One factor is the normal variation in either the thickness or density of the human body, so that there is a wide range of exposure values reaching the film within an individual radiograph. A chest film is a good example. The lung areas are not so dense, and the film exposure for these areas is relatively high. On the other hand, the mediastinum and spine are quite dense and produce a relatively low film exposure. In a situation like this it is not possible for all areas within the radiograph to receive the exposure value that would produce maximum contrast. We will come back to this problem later. The second factor is when there is an error in setting the exposure factors so that the radiograph is either too light (underexposed) or too dark (overexposed). Both conditions reduce contrast.

All films have a limited exposure range in which they can produce contrast; if areas of a film receive exposures either below or above the useful exposure range, contrast will be diminished, or perhaps absent. Image contrast is reduced when a film is either under- or overexposed.

In our experiment we have observed that two characteristics of film, density and contrast, both change as the amount of exposure that the film receives is increased. However, the two characteristics change in different ways. The density increases continuously until it reaches a maximum value at the very high exposure levels. But contrast is different: It increases as the exposure is increased, but it reaches a maximum in the middle of the exposure range. If the exposure is increased beyond that point, the contrast decreases. At high exposure values the contrast will decrease to zero.

The relationship of density and contrast to exposure is often shown by means of graphs, which we should now become familiar with.

CHARACTERISTIC CURVE

If we take the *density* and the *relative exposure* values from our experiment and plot them on a graph, it would have the shape shown in Figure 15-3. This type of graph is known as either a film *characteristic curve* or an *H and D* (Hurter and Driffield) *curve*. The precise shape of the curve depends on the characteristics of the emulsion and the processing conditions. Different types of radiographic film have different curve shapes that change if the processing is changed, as discussed. The primary use of a characteristic curve is to show us the amount of contrast a film produces throughout a wide exposure range. At any exposure value, the contrast of the film is represented by the slope of the curve. At any particular point, the slope represents the density difference (film contrast) produced by a specific exposure difference (exposure contrast). A specific interval on the relative exposure scale represents the amount of exposure contrast delivered to the film during the exposure process. An interval along the density scale represents the amount of contrast that actually appears in the film. The slope of the characteristic curve at any point is the contrast factor because the contrast factor is the density difference (contrast) produced by a 2 to 1 exposure ratio (50% exposure contrast).

A film characteristic curve has three distinct regions, each with different contrast transfer characteristics. The part of the curve associated with relatively low exposures is designated the *toe*. It corresponds to the light or low density portions of an image. When an image is exposed so that areas fall within the toe region, little or no contrast is transferred to the image. In the film shown in Figure 15-2, the areas on the left correspond to the toe of the characteristic curve.

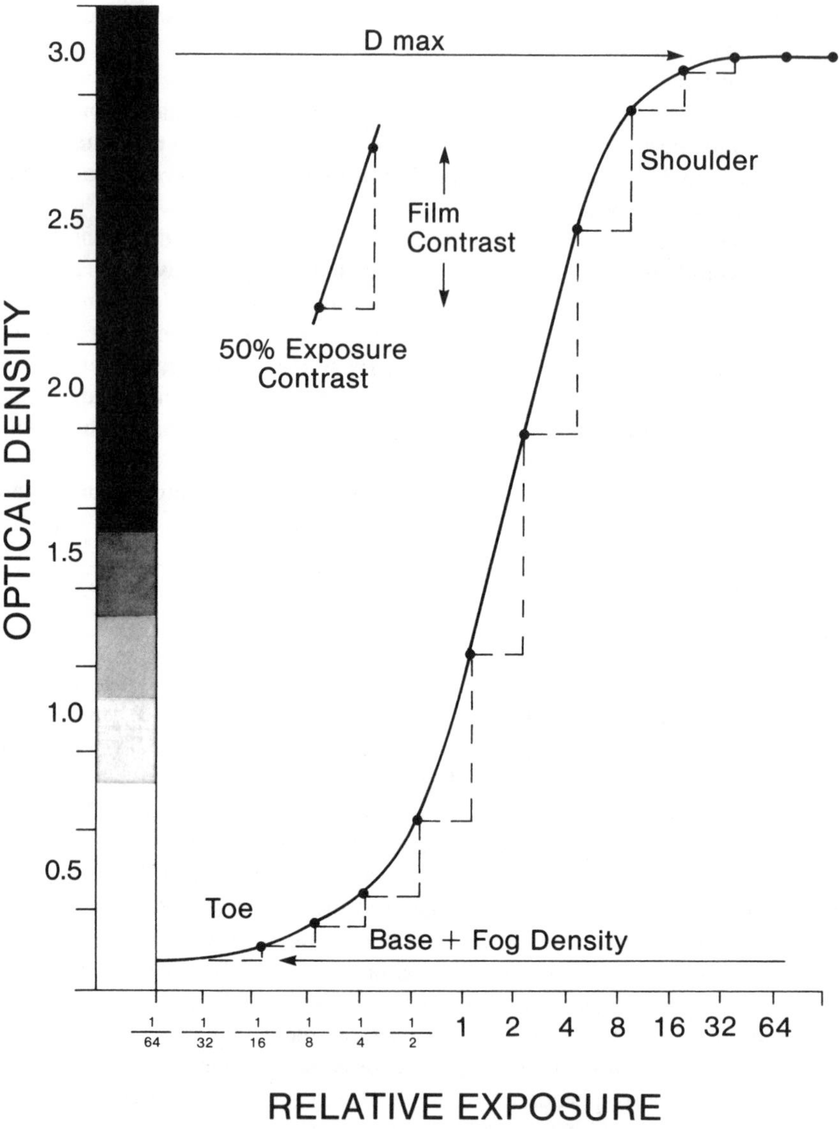

Figure 15-3 A film characteristic curve showing the relationship between density and relative exposure.

A film also has a reduced ability to transfer contrast in areas that receive relatively high exposures. This condition corresponds to the upper portion of the characteristic curve, where the slope decreases with increasing exposure. This portion of the curve is traditionally referred to as the *shoulder*. In Figure 15-2 the dark areas on the right correspond to the shoulder of the characteristic curve. The two significant characteristics of image areas receiving exposure within this range are that the film is quite dark (dense) and contrast is reduced. In many instances, some image contrast is present, but it cannot be observed on the conventional view box because of the high film density. This contrast can be made visible by viewing the film with the bright "hotlight."

The highest level of contrast is produced within a range of exposures falling between the toe and the shoulder. This portion of the curve is relatively straight and has a very steep slope in comparison to the toe and shoulder regions. In most imaging applications, it is desirable to expose the film within this range so as to obtain maximum contrast.

The minimum density, in the toe, is the base plus fog density. It is observed when unexposed film is processed and is typically in the range of 0.1 to 0.2 density units. The maximum density, in the shoulder, determined by the design of the film emulsion and the processing conditions, is typically referred to as the D_{max}.

Contrast Curve

It is easier to see the relationship between film contrast and exposure by using a contrast curve, as shown in Figure 15-4. The contrast curve shows the value of the contrast factor at different exposure values. We can plot a contrast curve by taking the contrast (density difference) between each exposure step shown in Figure 15-2 and plotting them against the relative exposure values.

The contrast curve corresponds to the slope of the characteristic curve. It clearly shows that the ability of a film to transfer exposure contrast into film contrast changes with exposure level, and that maximum contrast is produced only within a limited exposure range.

Since the contrast that a film produces changes with exposure, a specific film can be described completely only by using either a characteristic curve or a contrast curve, as illustrated in Figure 15-4. There are occasions, however, when it would be desirable to use a single-factor value to describe the general contrast characteristics of a film. Two different factors are used for this purpose: The *average gradient* expresses the average contrast transferring ability, and the *gamma* expresses the maximum contrast. These two factors do not give us a complete description of a film's contrast characteristic and have a limited use in radiography. The only complete and direct description is provided by the *contrast curve*, which shows us the contrast factor at each level of exposure.

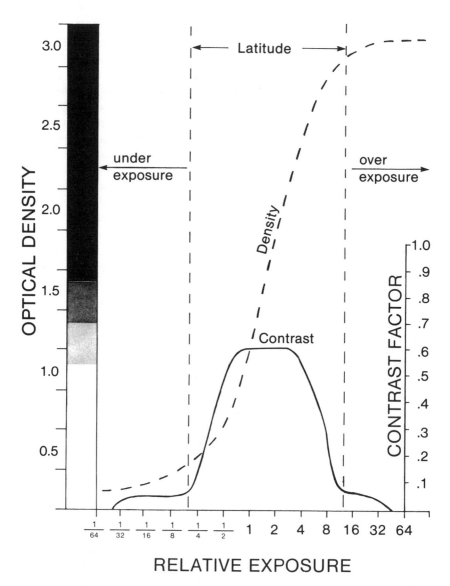

Figure 15-4 The relationship of film contrast (solid line) to relative exposure and the characteristic curve (dotted line).

FILM LATITUDE

The exposure range over which a film produces useful contrast is designated its *latitude*. The latitude is located around the center of the exposure scale, as shown in Figure 15-4. An underexposed film area below the latitude range contains little or no image contrast. Exposure values above the latitude range also produce areas with very little contrast and have the added disadvantage of being very dark or dense.

The latitude of a specific film is determined primarily by the composition of the emulsion and, to a lesser extent, by processing conditions. The significance of film latitude is that it represents the limitations of the exposure range that will yield useful image contrast.

The exposure to any given area of a film falls within one of three general ranges, as shown in Figure 15-4. Two general conditions can cause film exposure to fall outside the latitude range: (1) an incorrect exposure setting (exposure error) of the equipment, which can produce either an underexposure or an overexposure, and (2) a body section with a large variation in thickness and density, which produces an image with a wide range of subject contrasts that exceed the latitude range.

Exposure Error

In every imaging procedure it is necessary to set the exposure to match the sensitivity (speed) of the film being used. This is not always an easy task.

Exposure error is generally a much more significant problem in radiography than in other imaging procedures. It is not always possible to predict the amount of x-ray exposure required in every procedure because of slight variations in body size and composition. In any radiographic practice, a significant number of films must be repeated because of exposure error.

Subject Contrast Range

When an x-ray beam passes through certain body areas, the penetration of the areas varies considerably because of differences in tissue thickness and composition. Under these conditions it is possible for the range of exposures from the patient's body (subject contrast range) to exceed the latitude of the film. This typically produces a high level of area contrast, as discussed in a previous chapter.

When the exposure to some image areas falls outside the film latitude, details within those areas are recorded with reduced contrast, as illustrated in Figure 15-5. Notice that objects located within very thick and very thin body sections do not show up in the image because they are located in areas outside the film

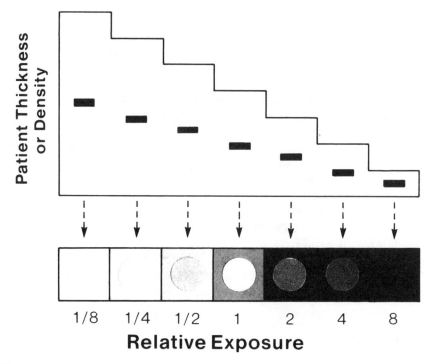

Figure 15-5 Loss of contrast in both thick and thin body sections when using high contrast film.

latitude. Radiography of the chest illustrates this problem: The area of the mediastinum receives a relatively low exposure, whereas the lung areas receive a much higher level.

One possible solution to the problem is to decrease the subject contrast range by using increased KV_p, compensation filters, bolus, or compression. Compensation filters are designed with different thicknesses across the x-ray beam. They are located so that the thicker section of the filter is over the thinner or less dense part of a patient's body. Another possible solution is to use a film with a longer latitude.

FILM TYPES

The overall contrast characteristic of a film (shape of its contrast curve and its latitude) is determined by the composition of the emulsion. Radiographic film is usually designated as either *high-contrast* or *medium-contrast* film. Medium-contrast film is often referred to as *latitude film*.

When selecting a film for a particular medical imaging application, contrast characteristics should be considered. Figure 15-6 compares the contrast charac-

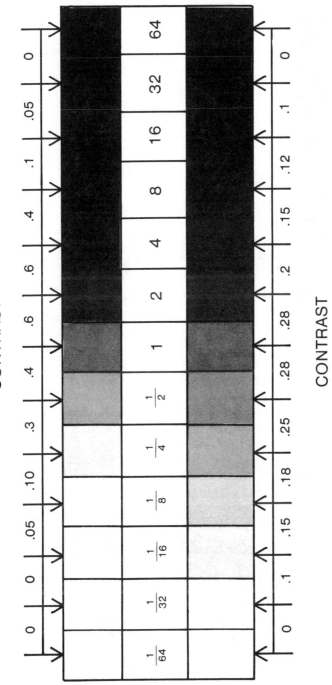

Figure 15-6 A comparison of high-contrast and medium-contrast or latitude film.

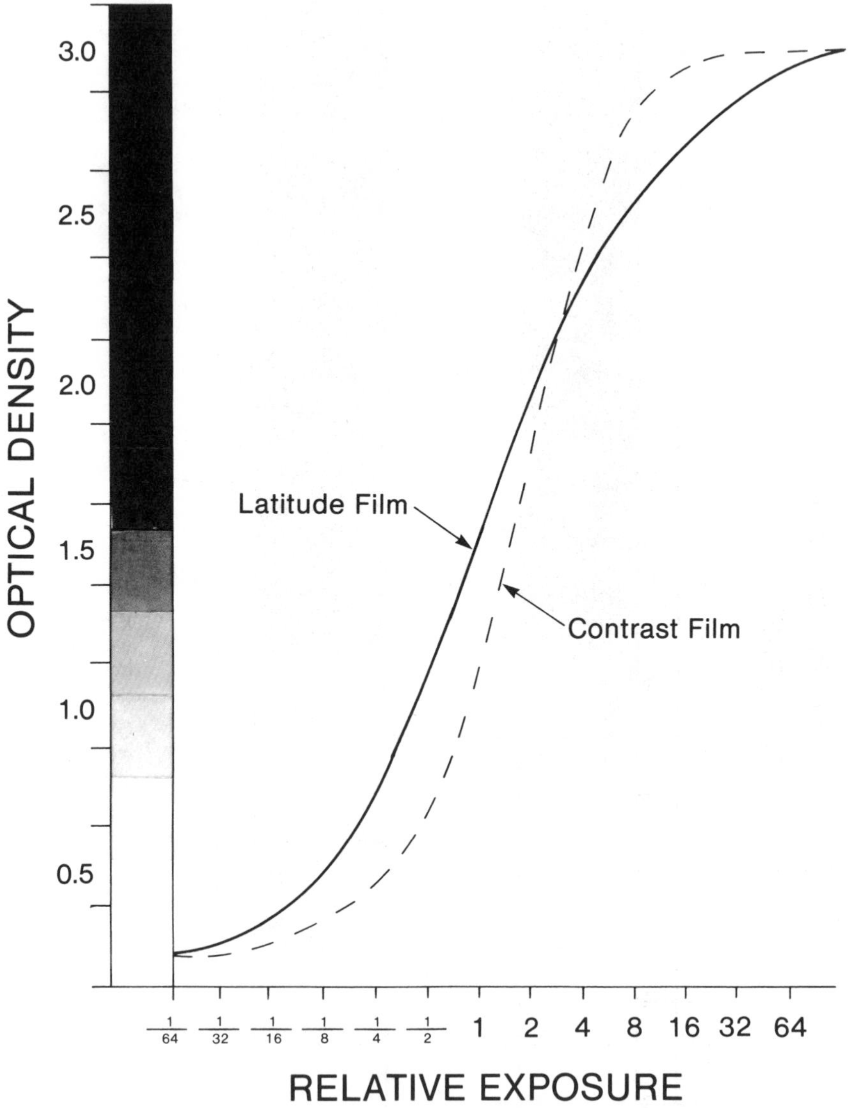

Figure 15-7 Characteristic curves for the high-contrast and latitude films illustrated in Figure 15-6.

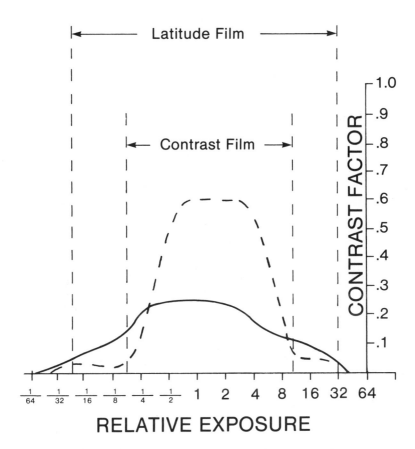

Figure 15-8 Contrast curves for the high-contrast and latitude films illustrated in Figure 15-6 (compare with characteristic curves in Figure 15-7).

teristics of two general types of radiographic film. High-contrast film can produce higher contrast. Notice the contrast of 0.6 between the areas with relative exposure values of 1 and 2. The contrast is limited, however, to a relatively small exposure range, or latitude. The medium-contrast (or latitude) film produces less contrast but can produce contrast over a much larger range of exposure values. The corresponding characteristic and contrast curves are shown in Figures 15-7 and 15-8.

Figure 15-9 illustrates how using a medium-contrast film actually increases object contrast within certain areas because of the overall reduction in area contrast.

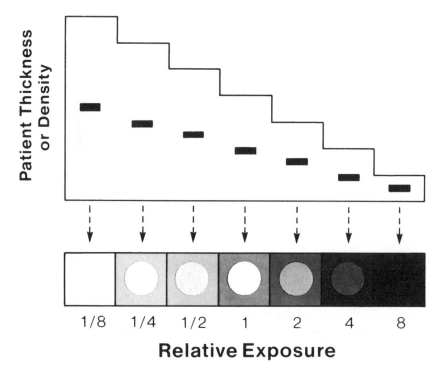

Figure 15-9 Increase in object contrast in thick and thin body sections with a latitude film (compare with Figure 15-5).

In the next chapter we will find that processing conditions can change the contrast characteristics of a film and must be controlled to give good image quality.

Study Activities

Explain how the characteristic curve displays the contrast characteristics of a film.

Define film contrast factor.

Explain the relationship between the contrast curve and the characteristic curve of film.

Explain film latitude and why it is important in radiography.

Identify the general conditions that make the use of a latitude film desirable.

Chapter 16

Film Processing

The creation of a visible radiographic image on film requires two steps, as shown in Figure 16-1. The first step is the *exposure*, which creates the invisible latent image, and the second step is the *processing*. Processing consists of passing the exposed film through a series of chemical solutions that cause the latent image to become visible. Processing is performed in an automatic film processor, which is relatively easy to use. You insert the film into the processor, and about 90 seconds later the processed and dry radiograph drops out ready to be viewed.

Although film processing is automatic, we need to know what goes on inside the processor and how it can affect the characteristics and quality of the radiograph. In this chapter we will learn what factors control the processing and how they can be selected and controlled.

One of the important functions of a radiographer is to make sure films are being properly processed. This requires a good knowledge of processing and of the quality control techniques that should be used.

THE PROCESSOR

When you are in an x-ray department you see how a processor looks on the outside. Let's now look inside, with Figure 16-2. This is a cross-section of a typical processor. It has three tanks of liquid: the developer, the fixer, and the washwater. The fourth section is where the film is dried by hot air before it drops out of the processor. The film is moved through the processor by the transport system, which consists of a series of rollers.

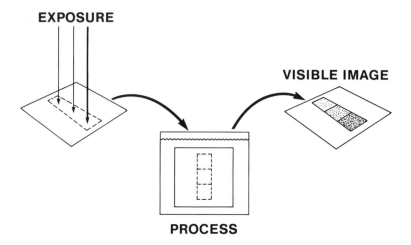

Figure 16-1 The two steps in the formation of a film image.

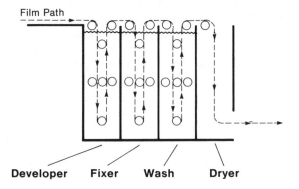

Figure 16-2 A film processor.

DEVELOPER

Development is a chemical process that causes the exposed film to turn dark (develop density). The *developer solution* supplies electrons that make their way into the sensitized grains and convert the other silver ions into black metallic silver. This causes the grains to become visible black specks in the emulsion. It is these many black specks that produce film density.

The developer is the first solution the film goes into. The developer consists of a mixture of chemicals. Each chemical performs a specific part of the development process.

When the developer solution is supplied to an x-ray facility, the different chemical components described below are already mixed together. However, developer is often supplied in a concentrated form that must be diluted with water before it is placed in the processor tank. The developer supplied by different manufacturers has the same basic chemical components described below, but they are not always mixed in the same way. Because of this, the developers supplied by different manufacturers will not necessarily produce films with identical characteristics. Changing from one brand of developer to another might change both the sensitivity and the contrast characteristics of a film.

We will now consider the basic chemical components found in a typical developer.

Reducer

Chemical reduction of the exposed silver bromide grains is the process that converts them into visible metallic silver. This action is typically provided by two chemicals in the solution: phenodin and hydroquinone. Phenodin is the more active of the two and primarily produces the mid to lower portion of the gray scale. Hydroquinone produces the very dense areas in an image.

Activator

The primary function of the activator, which is typically sodium carbonate, is to soften and swell the emulsion so that the reducers can reach the exposed grains.

Restrainer

Potassium bromide is generally used as a restrainer. Its function is to moderate the rate of development.

Preservative

Sodium sulfite, a typical preservative, helps protect the reducing agents from oxidation due to contact with air. It also reacts with oxidation products to reduce their activity.

Hardener

Glutaraldehyde is used as a hardener to retard the swelling of the emulsion. This is necessary in automatic processors in which the film is transported by a system of rollers.

AMOUNT OF DEVELOPMENT

In order for a film to produce the best possible image quality, it must receive just the right amount of development.

Underdevelopment

If a film does not receive enough development, some of the exposed grains will not be converted into visible density. The film will be lighter (less dense) than it should be. An underprocessed film often looks like an *underexposed* film. In fact, it is. This is because when a film is underdeveloped, it loses some of its sensitivity. Therefore, a normal exposure is not adequate and would result in an underexposed radiograph.

Underdevelopment also can cause a radiograph to have reduced contrast.

Overdevelopment

If a film is overdeveloped it will be darker (more dense) than it should be. It will look like an *overexposed* film. This is because overdevelopment will cause some of the unexposed grains to turn black. This has the effect of increasing the sensitivity of the film. When a film is overdeveloped, it takes less exposure to produce a certain density. However, this cannot be used as a method for reducing patient exposure because overdevelopment has several undesirable effects on the film. One is the increase in density of the light areas in a radiograph. This is actually a form of fog produced by the overdevelopment.

When a film is overdeveloped, it can also lose some of its contrast.

The amount of development a film receives is determined by the six factors we will now discuss.

Developer Composition

The processing chemistry of developer solutions supplied by different manufacturers is not the same. It is usually possible to process a film in a variety of developer solutions, but they will not all produce the same film sensitivity or contrast. The variation in sensitivity is usually relatively small but it must be considered when changing from one brand of developer to another.

Developer Concentration

Developer is usually supplied to a clinical facility in the form of a concentrate that must be diluted with water before it is pumped into the processor. Mixing errors that result in an incorrect concentration can produce undesirable changes in film sensitivity.

The test used to monitor developer concentration is the measurement of its *specific gravity*. The specific gravity of pure water has a value of 1. When chemicals are added to the water, this value increases in proportion to the concentration of the chemicals. The manufacturers of the developer will provide information on the correct specific gravity for a diluted solution.

Developer Replenishment

The film development process consumes some of the developer solution and causes the solution to become less active. Unless the solution is replaced, film sensitivity will gradually decrease.

In radiographic film processors, the addition of new developer solution, or its *replenishment*, is automatic. When a sheet of film enters the processor, it activates a switch that causes fresh solution to be pumped into the development tank. The replenishment rate can be monitored by means of flow meters mounted in the processor. The appropriate replenishment rate depends on the size of the films being processed. A processor used only for chest films generally requires a higher replenishment rate than one used for smaller films.

The replenishment rate can be adjusted and should be set either to values recommended by the manufacturer or to those that give correct processing as determined by quality control procedures.

Developer Contamination

If the developer solution becomes contaminated with another chemical, such as the fixer solution, abrupt changes in film sensitivity can occur in the form of either an increase or a decrease in sensitivity, depending on the type and amount of contamination. Developer contamination is most likely to occur when the film transport rollers are removed or replaced.

Development Time

When an exposed film enters the developer solution, development is not instantaneous. It is a gradual process during which more and more film grains are developed, resulting in increased film density. The development process is terminated by removing the film from the developer and placing it in the fixer. To some extent, increasing development time increases film sensitivity, since less exposure is required to produce a specific film density. In radiographic film processors, the development time is usually fixed and is approximately 20 to 25 seconds.

Development Temperature

The activity of the developer changes with temperature. An increase in temperature speeds up the development process and increases film sensitivity because less exposure is required to produce a specific film density.

The temperature of the developer in an automatic processor is controlled by a thermostat. It is usually set within the range of 90 to 95 °F. Specific processing temperatures are usually specified by film manufacturers.

There are several abnormal conditions that can cause an undesirable change in developer temperature. The thermostat or heater can malfunction. Also, if the temperature of the water flowing into the processor is not correctly regulated, it will have an effect on the developer temperature.

Developer temperature should be checked periodically to ensure that it is at the correct value.

FIXING

After leaving the developer, the film is transported into a second tank that contains the fixer solution. The fixer is a mixture of several chemicals that perform the following functions.

Neutralizer

When a film is removed from the developer solution, development continues because of the solution soaked up by the emulsion. It is necessary to stop this action to prevent overdevelopment and fogging of the film. Acetic acid is in the fixer solution used for this purpose.

Clearing

The fixer solution also clears the undeveloped silver halide grains from the film. Ammonium or sodium thiosulfate is used for this purpose. The unexposed grains leave the film and dissolve in the fixer solution. The silver that accumulates in the fixer during the clearing activity can be recovered; the usual method is to electroplate it onto a metallic surface within the silver recovery unit. Electroplating is a process in which an electrical current is run between two electrodes placed in a solution. The current picks up the silver ions in the solution and plates them onto one of the electrodes as metallic silver.

Preservative

Sodium sulfite is used in the fixer as a preservative.

Hardener

Aluminum chloride is typically used as a hardener. Its primary function is to shrink and harden the emulsion.

WASH

Film is next passed through a waterbath to wash the fixer solution out of the emulsion. It is especially important to remove the thiosulfate. If thiosulfate, also known as hypo, is retained in the emulsion, it will eventually react with the silver nitrate and air to form silver sulfate, a yellowish brown stain. The amount of thiosulfate retained in the emulsion determines the useful lifetime of a processed film. The American National Standard Institute recommends a maximum retention of 30 $\mu g/in^2$.

DRY

The final step in processing is to dry the film by passing it through a chamber in which hot air is circulating.

QUALITY CONTROL

Variations in processing conditions can produce significant differences in film sensitivity. One objective of a quality control program is to reduce exposure errors that cause either underexposed or overexposed film. Processors should be checked several times each week to detect changes in processing. This is done by exposing a test film to a fixed amount of light exposure in a sensitometer, running the film through the processor, and then measuring its density with a densitometer. It is not necessary to measure the density of all exposure steps. Only a few exposure steps are selected, as shown in Figure 16-3, to give the information required for processor quality control. The density values are recorded on a chart (Figure 16-4) so that fluctuations can be easily detected.

Base Plus Fog Density

One density measurement is made in an area that receives no exposure. This is a measure of the base plus fog density. A low density value is desirable. An increase in the base plus fog density can be caused by overprocessing a film.

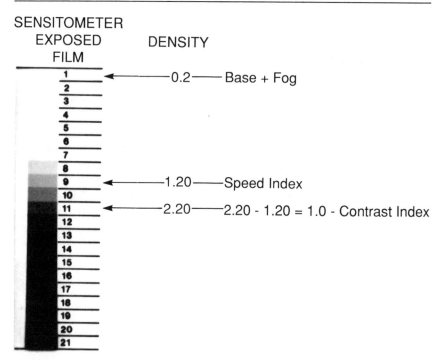

Figure 16-3 Density values from a sensitometer exposed film strip used for processor quality control.

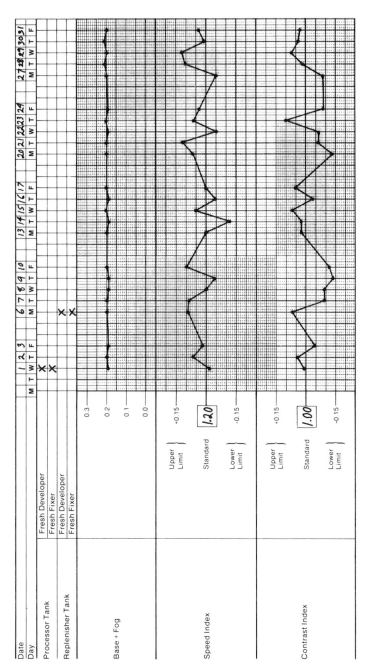

Figure 16-4 A processor quality control chart.

Speed

A single exposure step that produces a film density of about 1 density unit (above the base plus fog value) is selected and designated the "speed step." The density of this same step is measured each day and recorded on the chart. The density of this step is a general indication of film sensitivity or speed. Abnormal variations can be caused by any of the factors affecting the amount of development.

Contrast

Two other steps are selected, and the difference between them is used as a measure of film contrast. This is the contrast index. If the two sensitometer steps that are selected represent a two-to-one exposure ratio (50% exposure contrast), the contrast index is the same as the contrast factor discussed earlier. This value is recorded on the chart to detect abnormal changes in film contrast produced by processing conditions.

If abnormal variations in film density are observed, all possible causes, such as developer temperature, solution replenishment rates, and contamination, should be evaluated.

If more than one processor is used for films from the same imaging device, the level of development by the different processes should be matched.

Study Activities

Describe the changes that occur in film as it passes through the developer solution.

Explain the general function of fixing solution.

Explain why it is necessary to wash a film at the end of the development cycle.

Name six factors associated with development that can alter the sensitivity of film.

Chapter 17

Intensifying Screens

Conventional radiographic receptors contain two major components, the film and intensifying screens. Both are contained in either the cassettes or in film changers. Intensifying screens, you recall, absorb the x-ray energy and convert it into light. It is the light that then actually exposes the film.

Intensifying screens are used because film is much more sensitive to light than it is to x-radiation. Therefore, a radiograph can be created with much less exposure by using intensifying screens than if the film were exposed directly by the x-ray beam.

Intensifying screens are made of relatively thin sheets of fluorescent material that are placed in direct contact with the film, as shown in Figure 17-1. Most radiographic receptors use two screens, one on each side of a double emulsion film.

Intensifying screens are designed with different characteristics that must be considered when selecting screens for a specific clinical procedure and when selecting exposure factors. Image detail is the quality characteristic most affected by the intensifying screen, while contrast and noise are most affected by the film. The sensitivity of the receptor is determined by characteristics of both the intensifying screen and the film.

SCREEN FUNCTION

Intensifying screens are made of fluorescent materials. A fluorescent material is one that absorbs one form of radiation and converts it into another. Perhaps you are familiar with fluorescent objects that appear to glow in the dark when exposed to ultraviolet ("black") light. This happens because the fluorescent material converts invisible ultraviolet light into visible colors. The common fluorescent light gets its name from the fluorescent material coated on the inside of the glass tube.

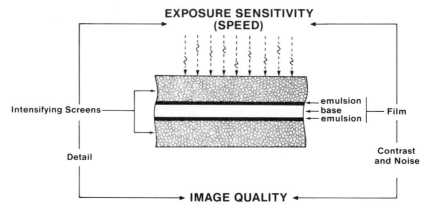

Figure 17-1 A conventional radiographic receptor.

Fluorescent light tubes also contain a gas that produces an almost invisible ultraviolet light when the electricity is turned on. The fluorescent coating converts the ultraviolet into the visible light which we see.

Fluorescent materials will also convert invisible x-radiation into visible light. This is what happens in the intensifying screen. Several different fluorescent compounds are used to make intensifying screens. It is not necessary to know these compounds, only their general characteristics. For many years calcium tungstate was the most common fluorescent compound used in intensifying screens. Today most screens are made from compounds that contain chemical elements belonging to the rare earth group. Their main advantage is that the rare earth elements are better x-ray absorbers than tungsten (in calcium tungstate), and this makes it possible to reduce patient exposure.

Figure 17-2 illustrates the basic function of an intensifying screen. When screens are used, a film exposure can be obtained with a very small fraction of the radiation that would be required if the film were exposed directly to x-radiation.

X-Ray Absorption

The first function of the intensifying screen is to absorb the x-ray energy that has passed through the patient's body. X-ray screens are generally good absorbers because they contain chemical elements that have good atomic numbers for this purpose. Unfortunately, most intensifying screens do not absorb all of the x-radiation. This radiation, which just passes through the screen, is wasted because it is not converted into light and used to expose the film. One of the characteristics of a screen that affects its absorption of x-radiation is its thickness.

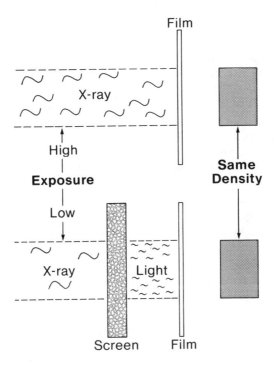

Figure 17-2 Conversion of x-ray energy in an intensifying screen.

Thick screens absorb more and waste less radiation than thin screens. However, as we will discover later, thick screens do not produce images with as much visibility of detail as thin screens. This is a factor to be considered when selecting screens for a specific clinical procedure.

Light Emission

The light that exposes the film is produced in the intensifying screens by the fluorescent process, which converts some of the x-ray energy into light energy. The thing we need to know about the light is its color. The color of the light is determined by the specific fluorescent compound used. Some screens are designed to produce blue light and some to produce green light. This must be taken into consideration when selecting film to use with a specific intensifying screen. Films are designed to be either blue-sensitive or green-sensitive. The film and the screen must be matched with respect to color.

SENSITIVITY

One of the important characteristics of an intensifying screen is its sensitivity, or the amount of radiation exposure it requires to produce an image. It is desirable for an intensifying screen to be highly sensitive and to require a low exposure. However, we will soon learn that the characteristics that make a screen more sensitive will also decrease the visibility of detail. When selecting an intensifying screen for a specific clinical procedure, we must consider this trade-off between sensitivity (radiation exposure) and visibility of detail.

The sensitivity value is the amount of exposure (in mR) needed to produce a film density of 1 unit above the base plus fog level.

In practice, we need to consider the exposure requirements (sensitivity) of the total image receptor, which includes both the intensifying screens and the film. This is determined by characteristics of both components. At this time we will consider the characteristics of an intensifying screen that affect overall receptor exposure.

Some manufacturers do not provide sensitivity values for their receptor systems, but most provide speed values such as 100, 200, 400, etc. The speed scale compares the relative exposure requirements of different receptor systems. Most speed numbers are referenced to a so-called *par speed system* that is assigned a speed value of 100. Whereas sensitivity is a precise receptor characteristic that describes the amount of exposure the receptor requires, speed is a less precise value used to compare film-screen combinations. There is, however, a general relationship between exposure requirements (sensitivity) and receptor speed values:

$$Sensitivity\ (mR) = 128/speed$$

For example, a receptor with a true speed value of 100 requires an exposure of 1.28 mR to produce a 1-unit film density. Sensitivity and speed values are inversely related. A more sensitive receptor has a higher speed value than a less sensitive receptor. The range of receptor sensitivity and speed values used in radiography is shown below:

Speed	Sensitivity (mR)
1200	0.1
800	0.16
400	0.32
200	0.64
100	1.28
50	2.56
25	5
12	10

Most receptors are given a nominal speed rating by the manufacturer. The actual speed varies, especially with KV_p and film processing conditions. The sensitivity of an intensifying screen-film receptor depends on the type of screen and film used, in addition to the conditions under which they are used and under which the film is processed.

We now consider characteristics of the screen that contribute to its sensitivity.

Photon Energy

The sensitivity of intensifying screens varies with x-ray photon energy because sensitivity is directly related to absorption efficiency. Absorption efficiency and screen sensitivity are highest when the x-ray photon energy is just above the K edge of the absorbing material. Each intensifying screen material generally has a different sensitivity-photon energy relationship because its K edge is at a different energy from the other materials.

The spectrum of photon energies within an x-ray beam is most directly affected and controlled by the KV_p; the sensitivity and speed of a specific intensifying screen are not constant but change with KV_p.

Significant exposure errors can occur if technical factors (KV_p and MAS) are not adjusted to compensate for variations in screen sensitivity. This often occurs when the same technique charts are used with screens composed of different materials. Also, the KV_p response characteristics of automatic exposure control sensors should be matched to those of the intensifying screens.

IMAGE DETAIL

The visibility of details of anatomy in a radiograph is affected by blurring, which has three possible sources: motion, the focal spot, and the receptor. The first two causes will be discussed in Chapter 18; let's consider the receptor now. Within the receptor, most of the blurring is produced by the intensifying screen. Figure 17-3 shows us how this happens.

Let's consider a small object like a calcification. The image of this object will be transferred to the intensifying screen by radiation moving along the dotted line. The light that creates the image will be produced along the path of the dotted line through the intensifying screen. Because the screen has some thickness, the light will spread over some area of it before it reaches the film. This is the blurring process. The amount of blurring is generally related to the thickness of the screen. A thick screen blurs the image more and reduces visibility of detail more than a thin screen.

Small Objects

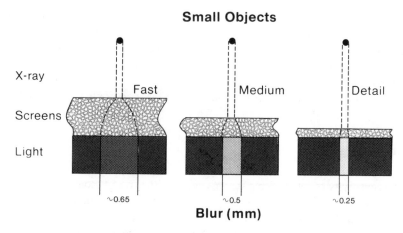

Figure 17-3 Effect of screen thickness on image blur.

This is a factor that must be considered when selecting intensifying screens for specific clinical applications. Thin screens produce images with better visibility of detail, but thin screens waste more radiation and have a lower sensitivity. They require a higher exposure. The most important thing to consider when selecting intensifying screens is the trade-off between image detail and exposure.

Screen-Film Contact

If the film and intensifying screen surfaces do not make good contact, the light will spread, as shown in Figure 17-4, and will produce image blurring. This abnormal condition occurs when a cassette or film changer is defective and does not apply enough pressure over the entire film area. Inadequate film-screen contact usually produces blurring in only a portion of the image area.

The conventional test for film-screen contact is to radiograph a wire mesh. Areas within the image where contact is inadequate will appear to have a different density than the other areas. This variation in image density is most readily seen when the film is viewed from a distance of approximately 10 feet and at an angle.

Crossover

If the film emulsion does not completely absorb the light from the intensifying screen, the unabsorbed light can pass through the film base and expose the emulsion on the other side. This is commonly referred to as *crossover*. As the light

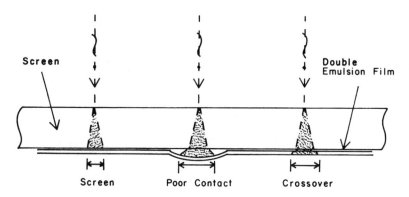

Figure 17-4 Sources of blur in screen-film receptors.

passes through the film base, it can spread and introduce image blur, as illustrated in Figure 17-4. Some films are designed to minimize crossover blurring. Crossover can be decreased by placing a light-absorbing layer between the film emulsion and the film base, using a base material that selectively absorbs the light wavelengths emitted by the intensifying screens, and by designing the film emulsion to increase light absorption.

Intensifying screens are usually identified by brand names, which do not always indicate specific characteristics. Most screens, however, are of five generic types:

1. mammographic
2. detail
3. par speed
4. medium speed
5. high speed

Figure 17-5 shows how these general screen types fit into the relationship between image blur and required exposure.

SCREEN TYPES

When selecting an intensifying screen for a specific clinical examination, the first consideration is image detail. If a radiograph does not show adequate detail, the accuracy of the diagnosis might be affected. The second factor to consider is patient exposure. You should not use a screen that gives more detail than is required because it will subject the patient to unnecessary exposure.

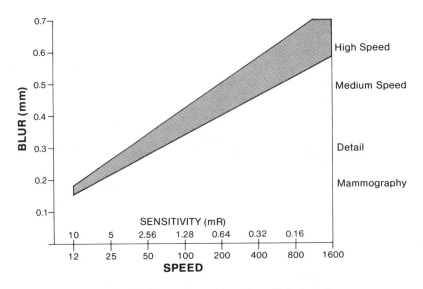

Figure 17-5 General relationship between image blur and sensitivity (speed).

Study Activities

Explain why intensifying screens are used.

Name the two basic functions performed by intensifying screens.

Identify the image quality characteristic that is most significantly affected by intensifying screens.

Explain the general relationship between image detail and the sensitivity (speed) of intensifying screens.

Describe the general conditions in which the use of a fast intensifying screen is appropriate.

Describe the general conditions in which the use of detail-intensifying screens is appropriate.

Chapter 18

Radiographic Detail

Of all medical imaging systems, radiography has the ability to produce images with the greatest detail. All radiographs, however, contain some *blur*, which reduces visibility of detail. The three basic sources of radiographic blurring and loss of detail are indicated in Figure 18-1: (1) the focal spot, (2) motion during film exposure, and (3) the receptor. Most receptor blur is produced by the spreading of light within the intensifying screen, as was described in Chapter 17. Two other sources of receptor blur are light crossover within the film base and poor contact between the intensifying screen and the film surface.

The amount of blur in a specific situation can be represented by a number value in millimeters. This value represents how much each point in the image is smeared or blurred. Visibility of detail is inversely related to the amount of blur. We have the best visibility of detail when blur values are low. The blur value is a reasonably good indicator of the smallest object that will be visible in a radiograph; we usually cannot see objects smaller than the blur value for a specific procedure.

The amount of blur in radiographs is generally in the range of 0.15 to approximately 1 mm. The *blur value* for a specific radiographic procedure depends on a combination of factors, including focal spot size, type of intensifying screen, location of the object being imaged, and exposure time (if motion is present). The general objective is not always to produce a radiograph with the greatest possible detail but to produce one with adequate detail without overheating the x-ray tube or overexposing the patient.

We begin by considering the characteristics of the three basic blur sources and then show how they combine to affect image quality.

OBJECT LOCATION AND MAGNIFICATION

Before proceeding with a discussion of the various types of blur, it is necessary to establish the relationship of several distances within the imaging system. Figure 18-2 shows the three basic distances. The focal-spot-to-receptor distance,

181

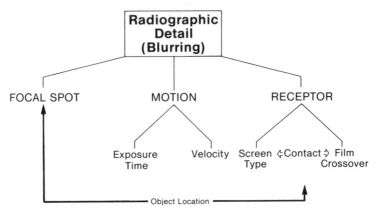

Figure 18-1 The sources of blur in radiographs.

FRD, is always the sum of the focal-spot-to-object distance, FOD, and the object-to-receptor distance, ORD. With respect to image formation and image quality, the significant values are not the actual distances but certain distance ratios.

In the formation of an x-ray image, the image will always be larger than the object if the object is separated from the receptor. The amount of enlargement, or magnification, is equal to the FRD to FOD ratio (the magnification factor). Magnification, m, is increased either by increasing the FRD or by bringing the object closer to the x-ray tube, which decreases the FOD.

Another useful relationship is the ORD to FRD ratio. This quantity, s, is used to specify the distance between the object and receptor (ORD) in relation to the total distance between focal spot and receptor (FRD). It is helpful to think of a scale running from the receptor to the focal spot: The receptor would be at the zero end and the focal spot would be at the other end, which always has a value of 1. The position of the object being radiographed can be specified with respect to this scale; for example, if it is located at $s = 0.2$ on the scale, the ORD is 20% of the FRD. For certain types of blurring, the blur value is dependent on s rather than on the actual distance between object and receptor.

When blur is given a specific value, the location within the imaging system must be specified. Blur values are generally specified for either the receptor location or the location of the object being radiographed. If an x-ray system has a blur value of 0.3 mm when measured at the location of the object and a magnification factor of 1.2, the image at the receptor location will have a blur value of 0.3×1.2 or 0.36 mm. On the other hand, if an imaging system has a blur value specified at the receptor, the amount of blur at the object is found by dividing by the magnification factor. In all cases, as shown in Figure 18-3, the relationship between the blur at the object and at the receptor surface is just the magnification, m.

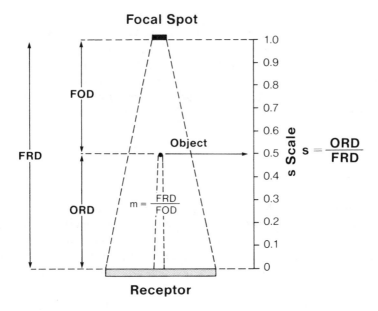

Figure 18-2 Distance relationships in radiographic imaging.

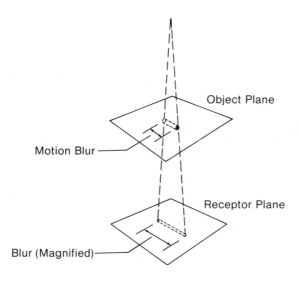

Figure 18-3 Magnification of blur.

The visibility of detail within a radiograph is determined by the relationship of the blur to the size of the objects being imaged. As blur values begin to exceed object dimensions, contrast and visibility are greatly diminished. Therefore, the most appropriate location for considering blur is at the location of the object where the blur value can be compared directly with object size or anatomical detail. Another major reason for considering blur at the location of the object rather than at the receptor is that the blur values from all contributing factors can be combined, as described below.

MOTION BLUR

Blurring will occur if the object being imaged moves during the exposure. The amount of blur is equal to the distance moved during the exposure, as shown in Figure 18-3. The effect of motion on each object within the body is to reduce its contrast by spreading the image over a larger area, as indicated. If motion of body parts cannot be temporarily halted, motion blur can be minimized by reducing exposure time.

A misconception regarding motion blur is that it is increased by magnification. Although it is true that the blur at the receptor surface is increased, image quality generally depends on the amount of blur at the location of the object. This value is not affected by magnification.

SIZE OF FOCAL SPOT BLUR

X-ray tube focal spots contribute to image blur. Consider the example shown in Figure 18-4. All of the x-ray photons that pass through each point of the object do not expose the film at one point. They are spread over an area, as indicated. This area is the blur produced by the size of the focal spot. The width of the area is the blur value, which depends on two things: (1) the size of the focal spot and (2) the location of the object along the s scale between the receptor and the focal spot. The blur value, B_f, with respect to the object size (at the object location) is given by

$$B_f = F \times s$$

where F is the dimension of the focal spot. Note that the value of focal spot blur, for a given focal size, is directly related to the position of the object on the s scale. If the object is in direct contact with the receptor ($s = 0$), focal-spot blur vanishes. As the object is moved away from the receptor, the blur increases in direct proportion to the value of s. The significance of this is illustrated in Figure 18-5.

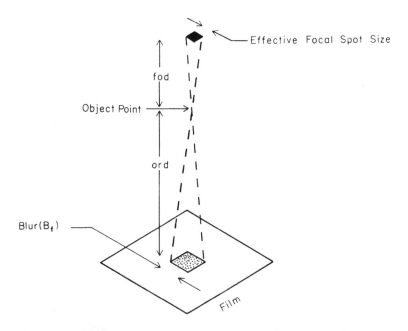

Figure 18-4 Focal spot blur.

When the object is moved away from the receptor, both the image and blur are magnified, but the blur increases faster than the size of the image. Therefore, the blur value is increased in proportion to the object size, causing a deterioration in image quality.

Figure 18-5 shows the relationship of focal-spot blur to object size as it is affected by object location(s). The amount of blur relative to the size of the object in the object plane increases in proportion to the relative distance(s) between the object and the receptor. The maximum blur occurs at the focal spot, where the blur value becomes equal to the actual size of the spot.

The effect of focal spot size on the visibility of a specific object depends on three factors: (1) the size of the object, (2) the size of the focal spot, and (3) the location of the object.

Focal Spot Size

The most significant characteristic of a focal spot is its size. Most x-ray tubes have labels that specify the size of the focal spot. The size specified by the

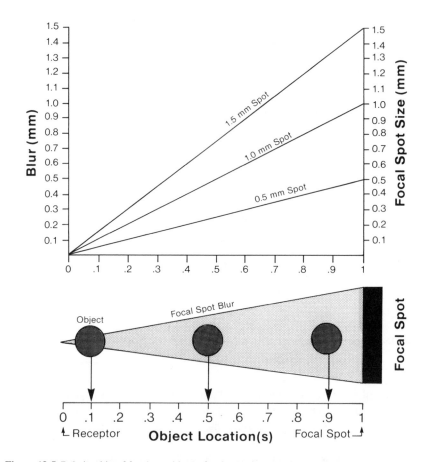

Figure 18-5 Relationship of focal spot blur to focal spot size and object location.

manufacturer on the label is generally referred to as the *nominal size*. The dimension that determines the blur characteristics of the focal spot is the *effective blur size*. The relationship between these two sizes is illustrated in Figure 18-6.

A distinction must be made between the focal spot dimension, which determines its blur characteristics, and the focal spot size, as specified by the tube manufacturer. Because of several factors, the generally stated size of a focal spot and its blur-producing size are significantly different. Blur is related to the effective size of a focal spot, rather than to its actual dimensions. The effective size is defined as the size of the focal spot if it had uniform x-ray emission over its surface rather than the usual nonuniform distribution discussed below.

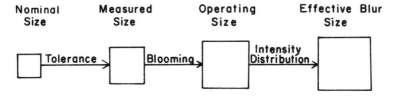

Figure 18-6 Different focal spot sizes.

Manufacturer's Tolerance

It is a general practice to allow a discrepancy between the manufacturer's stated nominal size and the actual measured size of focal spots. In almost all cases, the allowed tolerance is such that the stated nominal size is less than the measured size. The allowed tolerance generally depends on the size of the focal spot but is as large as 50% for the smaller focal spot sizes. For example, a focal spot that has a measured size of 0.9 mm could be labeled as a 0.6-mm focal spot because it falls within the accepted tolerance values.

Blooming

A common characteristic of many focal spots is that they undergo a change in size with changes in MA and KV_p. This effect is known as *blooming*. The size of a focal spot is relatively small at a low tube current but becomes larger at higher tube current values. The amount of blooming with an increase in tube current varies from tube to tube.

KV_p generally has less effect on focal spot size than current (MA). Some focal spots undergo a slight reduction in size with increased KV_p.

Intensity Distribution

Most focal spots do not have a uniform distribution of radiation over their entire area. Some points within the focal spot area produce more intense radiation than others. This nonuniform distribution of x-ray intensity causes a focal spot to have an effective blur size different from its actual physical size. The variation in x-ray emission across focal spots can be compared by the intensity profiles shown in Figure 18-7. The focal spot with a rectangular intensity distribution (center) has an effective blur size identical with the dimensions of the spot. A focal spot with a

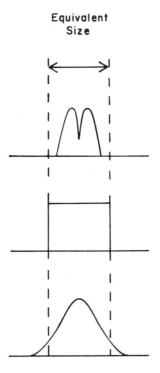

Figure 18-7 Three focal-spot intensity distributions with approximately the same effective blur size.

double peak distribution (top) has an effective blur size significantly larger than the actual dimension of the spot. This double-peak distribution is characteristic of many focal spots produced by conventional x-ray tubes.

The intensity distribution shown at the bottom of the illustration has a bell shape. A focal spot with this type of intensity distribution has an effective blur size less than its actual physical size. Bell-shaped focal spots are highly desirable because they have a relatively low effective blur size in comparison to their actual physical size and heat capacity. The three focal spots shown in Figure 18-7 are equal with respect to their effective blur size. However, they have different physical dimensions.

Anode Angle

The size of the focal spot is usually specified with reference to the center (central ray) of the x-ray beam area, or field. Because the anode surface is at an angle, the effective focal spot size changes with position in the field: It becomes smaller for points in the image toward the anode end and larger toward the cathode end of the tube. Let's now try a simple experiment to demonstrate this principle. Start by holding a sheet of notebook paper at arm's length in front of you so that you are looking directly at its surface. Now rotate the paper by bringing the bottom edge closer to you. Notice that as the angle changes the area that you see appears to decrease. When the paper becomes exactly horizontal you can't see any area, only the edge of the paper. In this experiment the sheet of paper represents the focal spot area on the surface of the anode. However, the effective size that is viewed by the radiographic receptor depends on the viewing angle. All points of the receptor do not view the anode surface at the same angle.

Measurement of Focal Spot Sizes

Two methods are used to determine the size of focal spots. The principles of the two methods are entirely different and generally produce different dimensions for the same focal spot. The *pinhole camera* is used by manufacturers to determine the actual dimensions of a focal spot, whereas the *star test pattern* is used in hospitals and clinics to determine the effective blur size.

Pinhole Camera

The principle of the pinhole camera is illustrated in Figure 18-8. The pinhole camera consists of a very small hole in a sheet of metal such as gold or lead. The pinhole is positioned between the focal spot and a film receptor, as shown. When the x-ray tube is energized, an image of the focal spot is projected through the pinhole onto the film. The size of the focal spot can be determined by measuring the size of the image and applying a correction factor if there is any geometric magnification present. If the pinhole is located at the midpoint between the focal spot and film, no correction factor will be required.

Star Test Pattern

The effective blur size of a focal spot can be measured by using a star test pattern, as shown in Figure 18-9. The first step in determining focal spot size is to make a radiograph with the test object located at approximately the midpoint between the focal spot and receptor. An image is obtained in which there is a zone of blurring surrounding the center of the star. The diameter of the blur zone is measured and used to calculate the size of the focal spot by using a formula, which is supplied with the star.

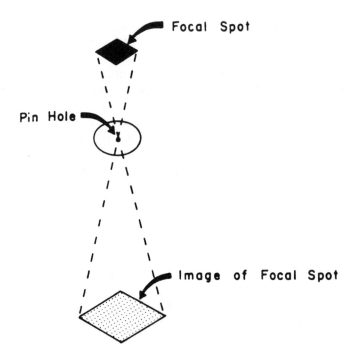

Figure 18-8 Pinhole method of determining focal spot size.

RECEPTOR BLUR

If the receptor input surface (intensifying screen) that absorbs the x-ray beam has significant thickness, blur will be introduced at this point. Blurring of this type generally occurs in intensifying screens and in the fluorescent layers of image intensifiers, to be discussed later. Blur production was illustrated in Figure 17-5. X-ray photons that pass through a point within the object are absorbed by the fluorescent layer and converted into light. The light created along the x-ray "path" spreads into the surrounding portion of the fluorescent layer. When the light emerges from the screen, it covers an area that is larger than the object point. In other words, the x-rays that pass through each point within an object form an image that is blurred into the surrounding area.

Because this type of blur is caused by the spreading (diffusing) of light, the blur profile generally has a shape different from that of motion blur. In most cases, the blur profile is bell-shaped. Because it is somewhat difficult to specify an exact blur

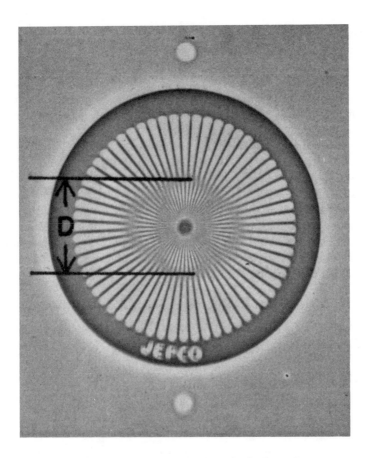

Figure 18-9 The image of a star test pattern used to determine focal spot size.

dimension, receptor blur is more appropriately described in terms of an *effective blur value*. The effective blur value of a receptor is similar in principle to the effective size of a focal spot. We can relate the dimension of the blur to image detail and also compare it to focal spot blurring. The amount of blur is primarily dependent on the thickness of the fluorescent layer.

Intensifying screens generally have effective blur values in the range of 0.15 to 0.6 mm. The approximate breakdown for the basic screen types is as follows:

- mammographic: 0.15 to 0.2 mm
- detail: 0.2 to 0.35 mm

- medium speed: 0.5 to 0.6 mm
- high speed: 0.6 to 0.7 mm

Since there must be a compromise between image detail and patient exposure, the objective, to repeat, is not always to use intensifying screens that produce maximum detail but to select screens that provide adequate detail with the lowest possible exposure.

In a given imaging system, the receptor blur value with respect to the size of the object can be decreased by introducing magnification. This is illustrated in Figure 18-10, in which a small object is being imaged. The presence of receptor blur produces a zone of unsharpness around the object. The actual blur dimension at the receptor surface is fixed by the receptor characteristics and is unaffected by magnification. When magnification is relatively small, the blur is large in comparison to the object. When magnification is increased, the relative blur-to-image size is decreased. What actually happens is that magnification causes the image of

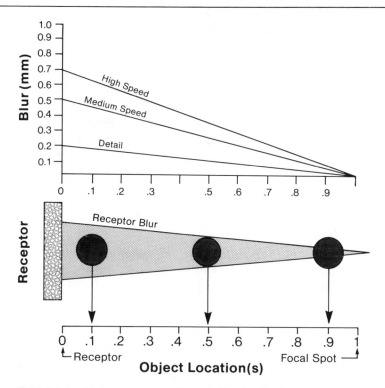

Figure 18-10 Relationship between receptor blur and object location for three types of intensifying screens.

a specific object to be enlarged at the receptor. This changes the relationship between the amount of blur and the size of the object *as it appears in the image* because the blur value at the receptor remains fixed and the object is magnified. However, the amount of receptor blur at the location of the object is reduced by magnification. Receptor blur at the object, Br, is related to the position of the object, s, by

$$Br = Bro(1 - s)$$

where Bro is the effective blur value of the receptor at the receptor surface. Although magnification can be used to reduce receptor blur with respect to object size, it must be used with caution. Since focal spot blur increases with magnification, the two blur sources must be considered together.

COMPOSITE BLUR

The total blur in an image is a composite of the three different types of blur: (1) motion, (2) receptor, and (3) focal spot. In some situations one of the three types of blur might predominate. When this occurs, the total composite blur of the radiographic system is essentially equal to the largest of the three blur values.

In order to determine the total, or composite, blur in a system, it is necessary to combine the three blur values.

The relationship between blur and object position is easy to visualize by using a blur nomogram, as shown in Figure 18-11. The nomogram has three scales: (1) blur, (2) focal spot size, and (3) the position of the object (s scale). The lines representing the blur from the three sources are drawn on the diagram according to the following simple rules:

1. The line representing receptor blur is drawn between a point on the blur scale that represents the effective blur of the particular receptor being used and a point located at a value of 1 on the s scale.
2. The line representing focal spot blur is drawn between the zero point on the s scale and a point on the focal spot scale that corresponds to the size of the focal spot.
3. The line for motion blur is drawn horizontally and intersects the blur scale at a value equal to the distance the object moved during the exposure interval.

The significance of the horizontal line for motion blur is that its value relative to the size of an object does not change with magnification. In most applications, the actual value for motion blur is difficult to estimate. It is included in this illustration primarily for the sake of completeness.

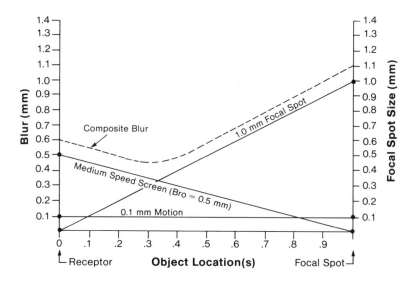

Figure 18-11 A nomogram used to determine radiographic blur at different object locations.

The three lines represent the blur of an object from the three blur sources. The composite, or total blur, is found for any point on the s scale by combining the three blur values, using a formula that need not be discussed here. It is of special significance that the total system blur usually has a minimum value at a point on the s scale. An exception is when the major source of blur is motion, in which case the total blur is essentially the same at any point along the s scale. However, when most of the blur arises from either the receptor or the focal spot, a minimum point is usually present. The position of the minimum blur point along the s scale depends on the characteristics of the receptor and the focal spot. If the receptor has a relatively low effective blur value, the minimum point will be located in the lower portion of the s scale, which represents a position close to the receptor surface. Either increasing receptor blur or decreasing focal spot size shifts the minimum blur point to higher values on the s scale. Inspection of the nomogram leads to two significant observations:

1. When an object is located near the receptor surface (low s-scale value), the total blur is essentially determined by the receptor.

2. As an object is moved away from the receptor surface (high s-scale value), the focal spot becomes the major determining factor in overall system blur.

Figure 18-12 shows the composite blur for a radiographic system using high-speed screens and a 0.5-mm focal spot. With this combination, the intensifying screen is the most significant blur source. Notice that the blur decreases with magnification over the useful range of object locations.

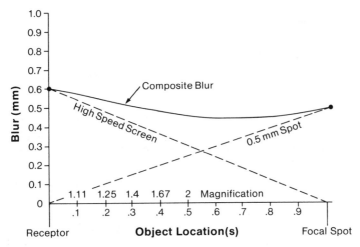

Figure 18-12 Blur produced by high-speed screens and a 0.5-mm focal spot.

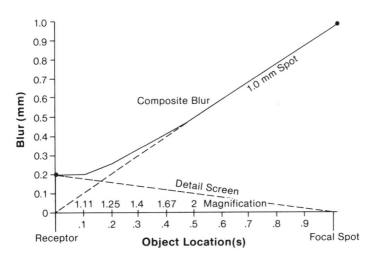

Figure 18-13 Blur produced by detail screens and a 1-mm focal spot.

Figure 18-13 illustrates the blur produced when using detail screens and a 1-mm focal spot. With this combination, the focal spot is the predominant blur source for most object locations.

These two examples show how either the receptor or the focal spot can be the predominant source of blur in a specific radiographic system. The dominating blur source is determined by the relationship of receptor blur to focal spot size and the location of the object; these are the factors that must be considered when setting up a radiographic system.

SUMMARY

Visibility of detail is reduced by blurring, which can come from three possible sources: motion, the focal spot, and the receptor. The amount of focal spot and receptor blurring depends on the location of the object being radiographed. If the object is close to the receptor, then the receptor becomes the most significant source of blurring. However, when the object is located away from the receptor and closer to the focal spot, receptor blurring decreases and the focal spot becomes the most significant source.

For a radiographic system with a specific receptor (intensifying screen type) and focal spot size, there is an object location that will produce the minimum blur and the best visibility of anatomical detail.

Study Activities

Name the three major sources of image blurring that reduce visibility of detail.

Calculate the amount of focal spot blurring for an object located halfway between a receptor and a 1.5-mm focal spot.

Calculate the amount of focal spot blurring if an object is located at 1/10th the distance (s = 0.1) to a 1.5-mm focal spot.

Explain briefly how the size of a focal spot is measured.

Describe the best place to locate an object to reduce the amount of focal-spot blurring.

Explain how the effect of receptor blurring is related to the location of the object being radiographed.

Chapter 19

Image Noise

It is generally desirable for film density, or image brightness, to be uniform except where it is changed by the body to form an image. There are factors, however, that tend to produce small variations in film density even when no image detail is present. This variation, called *image noise*, is usually random and has no particular pattern. It reduces image quality and is especially significant when the objects being imaged are small and have relatively low contrast.

All medical images contain some visual noise. The presence of noise gives an image a mottled, grainy, textured, or snowy appearance. In fact, several of these names are often used to describe image noise. In x-ray imaging you may hear the name *mottle*, which has been used for many years. However, noise is the preferred name today. The "snow" that we see on television is another form of image noise. A grainy photograph is a noisy image. Figure 19-1 compares images with different levels of noise. Image noise comes from a variety of sources, as we will soon discover. No imaging method is free of noise, but noise is much more common in certain types of imaging procedures than in others.

Nuclear medicine images are generally the most noisy. Noise is also significant in magnetic resonance imaging (MRI), computed tomography (CT), and ultrasound imaging. In comparison to these modalities, radiography produces images with the least noise. Fluoroscopic images are slightly more noisy than radiographic images, for reasons to be explained later. Conventional photography produces relatively noise-free images except where the grain of the film becomes visible.

In this chapter we consider some of the general characteristics of image noise, along with the specific factors in radiography and fluoroscopy that affect noise.

EFFECT ON VISIBILITY

Although noise gives an image a generally undesirable appearance, the most significant factor is that noise can cover up and reduce the visibility of certain

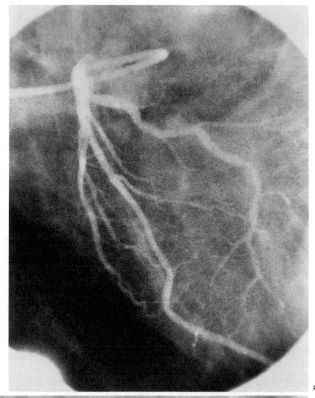

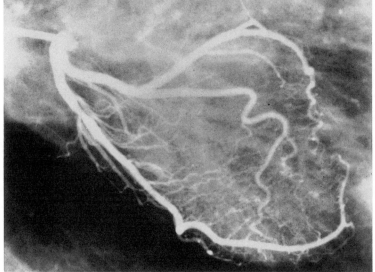

Figure 19-1 The image on the right (*B*) has more noise than the image on the left (*A*).

features within the image. The loss of visibility is especially significant for low-contrast objects. The visibility threshold, especially for low-contrast objects, is very noise-dependent. In principle, when we reduce image noise, the "curtain" is raised somewhat, and more of the low-contrast objects within the body become visible.

If the noise level can be adjusted for a specific imaging procedure, then why not reduce it to its lowest possible level for maximum visibility? Although it is true that we can usually change imaging factors to reduce noise, we must compromise between achieving the best image quality and patient exposure to radiation. There are also compromises between noise and other image characteristics, such as contrast and blur or visibility of detail. In principle, the user of each imaging method must determine the acceptable level of noise for a specific procedure and then select imaging factors that will achieve it with minimum patient exposure or effect on other image quality characteristics.

QUANTUM NOISE

X-ray photons strike a surface in a random pattern. No force can cause them to be evenly distributed over the surface. One area of the receptor surface might receive more photons than another area, even when both are exposed to the same average x-ray intensity.

In all imaging procedures using x-ray photons, most of the image noise is produced by the random manner in which the photons are distributed within the image. This is generally termed *quantum noise*. Recall that each individual photon is a quantum (specific quantity) of energy. It is the quantum structure of an x-ray beam that creates quantum noise.

Let us use Figure 19-2 to refresh our memories about the quantum nature of radiation. Here we see the part of an x-ray beam that forms the exposure to one small area within an image. Remember that an x-ray beam is a shower of individual photons. Because the photons are independent of one another, they randomly fall within an image area, somewhat like the first few drops of rain falling on the ground. At some points there might be clusters of several photons (drops) while at others there may be only a few. This uneven distribution of photons shows up in the image as noise. The amount of noise is determined by the variation in photon distribution from point to point within a small image area.

If we could cause the photons to be more evenly distributed in an area, this would reduce the amount of noise. In fact, there is a way to do this. Any time we increase the exposure or number of photons in an area, they will become more evenly distributed. This is demonstrated by the two areas in the lower portion of Figure 19-2. The areas on the left received an average exposure of 100 photons. However, the variation from area to area is about 10%. In an image this variation

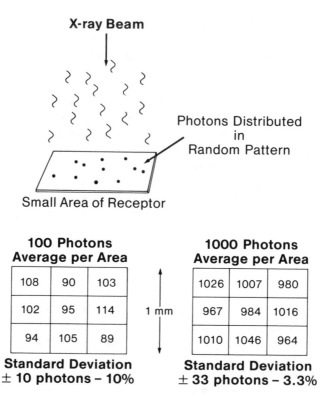

Figure 19-2 The concept of quantum noise.

would appear as noise. If the exposure were increased to an average of 1,000 photons, they would be more uniformly distributed, as shown on the right. Here the variation from area to area is only 3.3%. The variation or noise is decreased by increasing the exposure to the receptor.

We have just observed what is perhaps the most important characteristic of quantum noise: It can be reduced by increasing the concentration of photons (ie, exposure) used to form an image. (More specifically, quantum noise is inversely proportional to the square root of the exposure to the receptor.)

The relationship between image noise and required exposure is one of the issues that must be considered by persons setting up specific x-ray procedures. In most situations patient exposure can be reduced, but only at the expense of increased quantum noise and thus, possibly, reduced visibility. It is also possible, in most situations, to decrease image noise, but a higher exposure is required. Most x-ray procedures are conducted at a point of reasonable compromise between these two very important factors.

RECEPTOR SENSITIVITY

The photon concentration, or exposure, that is required to form an image is determined by the sensitivity (speed) of the receptor. The sensitivities of the receptors used in radiography and fluoroscopy vary over a considerable range, as shown in Figure 19-3. This chart shows the approximate values used for specific imaging applications.

Since quantum noise is related to receptor exposure, it is obvious that there is a considerable range of noise in x-ray images. Images created with a low receptor exposure, like fluoroscopy, will have much more noise than most other images. Thirty-five-mm cine films, used in cardiac imaging, are generally more noisy than conventional radiographs. A radiograph made with a 1,200-speed system will be much noisier than one made with a 50-speed system. Even though the amount of noise in an image is basically related to the total sensitivity of the receptor, there are some additional factors that have an influence on the amount of noise.

Screen-Film Radiography

The sensitivity of a radiographic receptor (cassette) is determined by characteristics of both the screen and the film and the way they are matched. The amount of image noise is affected differently if you change the sensitivity by changing the intensifying screen or film.

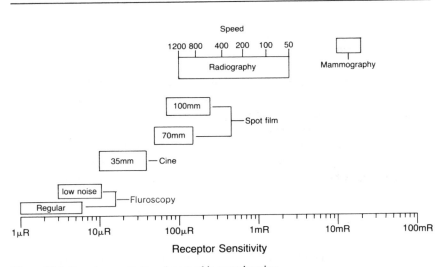

Figure 19-3 Receptor sensitivity values used in x-ray imaging.

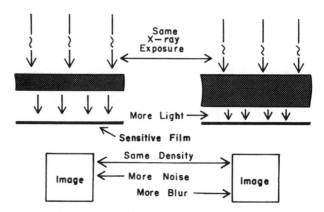

Figure 19-4 Comparison of image quality between two screen-film combinations.

Two screen-film combinations with the same total sensitivity are shown in Figure 19-4. One system uses a relatively thick high-speed screen and a film with conventional sensitivity. The other system uses a thinner detail-speed screen and a more sensitive film. The images produced by these two systems differ in two respects. The system using the thicker screen has more blur but less quantum noise than the system using the more sensitive film and thin screen. The reduction in noise comes from increased absorption efficiency and blur. The moral of this story is that you cannot completely predict the noise in an image by considering only the total receptor sensitivity. You must take into account the characteristics that are producing the sensitivity.

Intensified Radiography

Quantum noise is sometimes more significant in intensified radiography (cine and spot films) than in screen-film radiography because of generally higher receptor sensitivity values (ie, lower receptor exposures). With such systems, the quantum noise level can be adjusted.

Changing the size of the optical aperture in the camera is the most common method of adjusting the receptor sensitivity and quantum noise level. This adjustment can usually be made in the clinical facility by a service engineer.

Fluoroscopy

The same basic principles apply to a fluoroscope imaging system except that the sensitivity of the video camera is generally not fixed but can be varied through adjustments. The quantum noise level for a fluoroscope is generally set to an acceptable level by adjusting either the video camera or the aperture or both. The receptor sensitivity of a conventional fluoroscope is typically in the range of 1 to 10 μR per image frame. This relatively low exposure produces images with considerable quantum noise. In normal fluoroscopic viewing, however, we do not see one image frame at a time but an average of several frames, as discussed below.

Some fluoroscopic systems can be switched into a low-noise mode, which will improve the visibility of low-contrast detail. In the low-noise mode, the receptor sensitivity is reduced, and more exposure is required to form the image.

It is possible to develop x-ray receptor systems that would have greater sensitivity and would require less exposure than those currently used in x-ray imaging. But there is no known way to overcome the fundamental limitation of quantum noise. The receptor must absorb an adequate concentration of x-ray photons to reduce noise to an acceptable level.

GRAIN AND STRUCTURE NOISE

Although the quantum structure of the x-ray beam is the most significant noise source in most x-ray imaging applications, the structure of the film, intensifying screens, or intensifier tube screens can introduce some noise into images.

Recall that an image recorded on film is composed of many opaque silver halide grains. The grains in radiographic film are quite small and are not generally visible when the film is viewed in the conventional manner. The grainy structure sometimes becomes visible when a radiograph is optically enlarged, as when projected onto a screen. Whenever it is visible, film grain is a form of image noise.

Film-grain noise is generally a more significant problem in photography than in radiography, especially in enlargements of photographs recorded on film with a relatively high sensitivity.

Image-intensifying screens and the screens of intensifier tubes used in fluoroscopy are actually layers of small crystals. An image is formed by the production of light (fluorescence) within each crystal. The crystalline structure of screens introduces a slight variation in light production from point to point within an image. This structure noise is relatively insignificant in most radiographic applications.

ELECTRONIC NOISE

Video images, seen in fluoroscopy, often contain noise coming from various electronic sources. Video image noise is often referred to as snow. Some of the electronic components that make up a video system itself can be sources of electronic noise. The noise is due to random electrical currents, often produced by thermal activity within the device. Other electrical devices, such as motors and fluorescent lights, and even natural phenomena within the atmosphere generate electrical noise that can be picked up by a video system.

The presence of noise in a video system becomes especially noticeable when the image signal is weak. This effect can be easily observed by tuning a television (video) receiver to a vacant channel or a channel with a weak signal. The presence of excessive electronic noise in a fluoroscopic image is often the result of a weak video signal because of system failure or misadjustment.

EFFECT OF CONTRAST ON NOISE

The noise in an image becomes more visible if the overall contrast transfer of the imaging system is increased. This must be considered when using image displays with adjustable contrast, such as some video monitors used in fluoroscopy, and the viewing window in CT, MRI, and other forms of digital images. High contrast film also increases the visibility of noise.

EFFECT OF BLUR ON NOISE

The visibility of image noise can often be reduced by blurring because noise has a rather finely detailed structure. The blurring of an image tends to blend each image point with its surrounding area; the effect is to smooth out the noise and make it less visible.

The use of image blurring to reduce the visibility of noise often involves a compromise, because the blurring can also reduce the visibility of useful image detail.

High-sensitivity intensifying screens generally produce images showing less quantum noise than detail screens, because they produce more image blur. The problem is that no screen gives both maximum noise suppression and visibility of detail.

IMAGE INTEGRATION

Integration is the process of combining or averaging a series of images over a period of time. Since most types of image noise have a random distribution with respect to time, the integration of images can be quite effective in smoothing an image and reducing its noise content. Integration, in effect, blurs an image with respect to time rather than with respect to space or area.

Integration requires the ability to store or remember images, at least for a short period of time. Several devices are used for image integration in medical imaging.

Human Vision

The human eye responds to an average light intensity over a period of approximately 0.2 seconds. So this natural integration (averaging) is especially helpful when viewing fluoroscopic images.

The conventional fluoroscopic display is a series of 30 individual video images each second. Each image is displayed for one thirtieth of a second. Because a relatively low receptor exposure (less than 5 μR) is used to form each individual image, the images are relatively noisy. However, since the eye does not "see" each individual image but an average of several images, the visibility of the noise is reduced. In effect, the eye is integrating approximately six video images at any particular time. The noise actually visible to the human eye is not determined by the receptor exposure for individual fluoroscopic images but by the total exposure for the series of integrated images.

Digital Processing

When a series of images is acquired and stored in a digital imaging system memory, the images can be averaged to reduce the noise. This process is frequently used in digital subtraction angiography (DSA) and MRI.

IMAGE SUBTRACTION

There are several radiography applications in which one image is subtracted from another. A specific example is DSA. A basic problem with any image subtraction procedure is that the noise level in the resulting image is *higher* than in either of the two original images. This occurs because of the random distribution of the noise within each image.

Relatively high exposures are used to create the original images in DSA. This partially compensates for the increase in noise produced by the subtraction.

Study Activities

Describe the general appearance of noise in a radiograph.

Explain how quantum noise is produced.

Explain how you as a radiographer can change the amount of quantum noise in a radiograph.

Explain why noise is generally undesirable in an x-ray image.

Identify the specific component in the radiographic system that most directly affects the amount of quantum noise in the image.

Explain why a fluoroscopy image generally contains more noise than a radiograph.

Chapter 20

Radiographic Density Control

Maximum visibility in a radiograph requires that the optical density be within a range that produces adequate contrast, as discussed in Chapter 15. This is achieved by setting the exposure to fit the conditions established by the receptor system and the patient. The exposure can be selected either by manually adjusting the KV, MA, and exposure time or by using the automatic exposure control (AEC) circuit of the x-ray machine. Neither method produces a perfect exposure each time. A number of sources of exposure error must be considered during the production of a radiograph.

A proper film density is obtained when the radiation exposure penetrating the patient's body (receptor exposure) matches the sensitivity requirements of the receptor system. Both receptor sensitivity and receptor exposure are affected by many factors, as discussed above and shown in Figure 20-1. Film density is optimal when all of these factors are properly balanced.

After a radiographic system is installed, the films, screens, and grids are selected, and the processor is adjusted, the major task is selecting KV and MAS values to compensate for variations in patient thickness and composition. If the KV and MAS are to be selected manually, technique charts should be used for reference. The most common chart form gives KV and MAS values in relation to the thickness of different parts of the body. It should be emphasized that a given technique chart should be used only if it has been calibrated for the specific machine and film-screen-grid combination being used.

Exposure errors are produced when any of the factors listed in Figure 20-1 are not properly compensated for. When it is necessary to change certain factors, such as focal-receptor distance (FRD) or KV, to meet a specific examination objective, the change can usually be compensated for by changing another factor according to established relationships, such as the inverse-square law and the 15% rule.

In this chapter we consider the specific factors that relate to exposure (density) control, exposure error, and technique compensation.

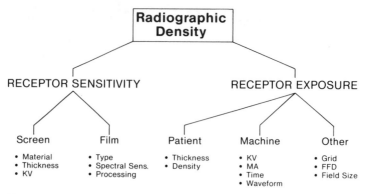

Figure 20-1 Factors that affect radiographic density.

THE X-RAY GENERATOR

The exposure delivered by an x-ray generator can be controlled by selecting appropriate values for MA, exposure time, and KV_p. In principle, several combinations of these three produce the same film density; therefore, other factors, such as patient exposure, x-ray tube heat production, generator capabilities, image contrast, and image blur, must be considered in setting technical factors.

MA

The intensity (exposure rate) of an x-ray beam is directly proportional to the MA value. The typical radiographic generator provides several MA values (25, 50, 100, 200, 500, etc) to choose from. The selection of an MA value is usually coupled with the selection of focal spot size. The use of the small focal spot (for better image detail) is typically limited to the lower MA values. The highest available MA value on a particular machine is determined by the capacity of the generator.

No one specific MA value must be used for a given procedure; the MA value must be selected in conjunction with the exposure time and KV_p. Some general rules governing the selection of MA are

- select low MA values to permit use of a small focal spot when image detail is important
- select large MA values to reduce exposure time when motion blurring is a problem
- select high MA values when it is desirable to reduce the KV_p to increase image contrast

Two types of exposure errors are associated with the MA selection: One is primarily a human error, and the other is an error associated with the calibration of the x-ray generator.

An exposure error can occur when the operator selects an MA value that is inappropriate in relation to the exposure time and KV_p value. This can occur if patient size and condition are not properly evaluated or if the technique charts are not correctly calibrated for the specific x-ray generator.

Exposure errors will occur if the output exposure rate of the x-ray machine is not proportional to the indicated MA value at a particular setting. It is not uncommon for the actual MA value to be different from the value indicated by the MA selector. This error source can be minimized by the periodic calibration of the x-ray generator. Calibration consists of measuring the x-ray exposure rate at each MA value that can be selected.

Exposure Time

Since the exposure produced by an x-ray tube is directly proportional to exposure time, it can be controlled by selecting an appropriate exposure time value. In radiography, exposure times are selected either by the operator, who sets the timer before initiating the exposure, or by the AEC circuit, which terminates the exposure after the selected exposure has reached the receptor.

As with MA, no one exposure time value is correct for a specific procedure. Remember that it is the combination of exposure time, MA, and KV_p that determines exposure. Some general rules for selecting an appropriate exposure time are

- select short exposure times to minimize motion blurring and improve image detail
- select longer exposure times when motion is not a problem and it is necessary to reduce either MA or KV_p.

Exposure errors can result from the selection of an inappropriate exposure time by the operator or from the failure of the generator to produce the exposure time indicated on the time selector.

X-ray machine timers should be calibrated periodically to determine if the machine produces an exposure that is proportional to the indicated exposure time. Timers can be calibrated by several methods, including the use of a spinning top, an electronic timer, or the measurement and comparison of actual exposure output produced by each timer setting.

Exposure errors encountered in the use of AEC are discussed later.

KV$_p$

Film exposure is more sensitive to changes in KV$_p$ than to changes in either MA or exposure time. This is because the KV$_p$ affects several independent factors that contribute to film exposure. In Chapter 7 we saw that x-ray beam intensity increased rapidly with an increase in KV$_p$. A good approximation is that doubling the KV$_p$ increases exposure from an x-ray tube by a factor of 4. In Chapter 10 we observed that the penetration of radiation through an object, such as a patient's body, increases with KV$_p$. The increase in both x-ray production and penetration with KV$_p$ means that a relatively small change in KV$_p$ produces a relatively large change in the exposure penetrating the patient's body and reaching the receptor. It should be recalled (Chapter 17) that the sensitivity of intensifying screens changes with KV$_p$. Both the direction and amount of change depend on the specific screen material.

A general rule of thumb in radiography is that a 15% increase in KV$_p$ will double the exposure to the film. In other words, it takes only a 15% increase in KV$_p$ to produce the same increase in film exposure as a 100% increase in either MA or exposure time. The 15% rule is useful for comparing change in KV$_p$ and MAS, but it by no means expresses a precise relationship.

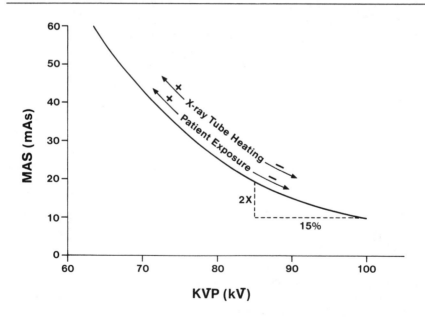

Figure 20-2 Relationship of KV$_p$ and MAS to control of film density.

Figure 20-2 shows the approximate relationship between MAS and KV_p values that would produce the same film exposure. The KV_p–MAS values represented by points along this curve apply to a specific x-ray generator, patient, and receptor system. If any of these three factors is changed, the position of the curve would be shifted.

Although it is true that KV_p-MAS values represented by any point along the curve produce the same film exposure, they will not produce the same image quality, patient exposure, and demands on the x-ray equipment. Moving down the curve (higher KV_p values) generally decreases patient exposure, x-ray tube heating, and motion blurring when the MAS is decreased by a shorter exposure time. The major reason for moving up the curve (higher MAS values) is to increase image contrast with the lower KV_p values, as discussed in Chapter 11.

The range of KV_p values for a specific procedure is selected on the basis of contrast requirements, patient exposure, and the limitations of the x-ray generator. However, small changes in KV_p within each range can be used to adjust film exposure.

Exposure errors can occur if the actual KV_p produced by an x-ray generator is different from the value indicated by the KV_p selector. Periodic calibration of an x-ray generator helps reduce this potential source of error.

Waveform

An x-ray generator that produces a relatively constant KV (ie, three-phase generator) requires less KV_p and/or MAS than a single-phase generator to produce the same film exposure. The constant potential, or three-phase, generator produces more radiation exposure per unit of MAS, as discussed in Chapter 8. For a specific KV_p value, the radiation is more penetrating because of the higher average KV during the exposure.

Technique charts designed for use with a single-phase generator would lead to considerable overexposure if used with a constant potential, or three-phase generator.

X-Ray Tubes

All x-ray tubes do not produce the same exposure (for a specific KV_p-MAS value), and the exposure output sometimes decreases with aging. A difference in tube output among tubes is often caused by variations in the amount of filtration. The significance of the tube-to-tube variation is that technique factors for one x-ray machine might not produce proper film exposure if used with another machine.

RECEPTOR SENSITIVITY

The overall sensitivity (speed) of the radiographic receptor is determined by the characteristics of both the film and the intensifying screens. The sensitivity of a specific film-screen combination is usually specified in terms of a speed value, as described in Chapter 17. If either the film or screen type is changed (old to new) so that the combined sensitivity changes, it will be necessary to change either the MAS or KV_p to compensate. If the speed is increased, less radiation will be required, so MAS or KV_p must be reduced. The relationship is

$$MAS\ (new) = (Speed\ (old)/Speed\ (new)) \times MAS\ (old)$$

For example, if we change from a 200- to 400-speed system,

$$MAS\ (new) = 200\ (old)/400\ (new) \times MAS\ (old) = 1/2 \times MAS\ (old)$$

A change in KV_p can also be selected to compensate for changes in receptors by using the 15% rule.

Many exposure errors are caused by undetected variations in receptor sensitivity from examination to examination, or over a period of time. When a receptor is described as having a specific speed value, such as 200, 400, etc, the value is nominal and applies to specific exposure and processing conditions. When these conditions change, so does the receptor sensitivity.

One of the major factors that produces variations in receptor sensitivity and therefore exposure error is variation in developer temperature and activity. The inherent sensitivity of film varies somewhat from one batch to another, but this is usually not sufficient to produce significant exposure error. Variations in screen sensitivity with KV_p can be a problem, especially when techniques appropriate for one type of screen (fluorescent material) are used for another type.

Grids

When a grid is changed, the exposure factors must be changed. The approximate relationship between the old and new MAS values depends on the Bucky factors, B, or the penetration values of the grids and is

$$MAS\ (new) = (B\ (new)/B\ (old)) \times MAS\ (old)$$

If one of the procedures being considered does not use a grid, then the value of the Bucky factor for it will be 1. The value of Bucky factors generally depends on grid ratio and the quantity of scattered radiation in the beam, as discussed in Chapter

12. Changing from a procedure without a grid to one with a grid Bucky factor of 5 requires the MAS to be increased by a factor of 5. Approximate Bucky factor values for different grids are given in Chapter 13.

PATIENT

For a given type of examination, the factor subject to the greatest variation from patient to patient is the penetration of the body section. For a given x-ray beam quality, or KV_p, the penetration depends on the thickness of the body part being examined and the composition of the body section. Changes in body thickness from one patient to another can be compensated for by changing either KV_p or MAS. An approximate relationship between KV_p and body thickness, t, is given by

$$KV_p = 50 + 2t \text{ (cm)}$$

For example, a 15-cm thickness would require a KV_p of approximately 80, whereas a 20-cm thickness would require a KV_p of 90.

When a change in patient thickness is compensated for by changing MAS, a change of a factor of 2 is generally required for a thickness difference of approximately 5 cm. This varies, however, with KV_p. A given thickness difference requires a smaller MAS change when higher KV_p values are used.

The presence of various pathological conditions can also alter body penetration and require appropriate exposure adjustments. Muscular patients generally require an additional increase in exposure factors, whereas elderly patients require a reduction.

DISTANCE AND AREA

As the area covered by the x-ray beam is increased, more scattered radiation is produced and contributes to film exposure. Although much of the scattered radiation is removed by the grid, it is often necessary to change exposure factors to get the same density with different field sizes.

Because of the inverse-square effect, the exposure that reaches the receptor is related to the focal spot-receptor distance, d. If this distance is changed (old to new), the new MAS value required to obtain the same film exposure will be given by

$$MAS \text{ (new)} = (d^2 \text{ (new)}/d^2 \text{ (old)}) \times MAS \text{ (old)}$$

A characteristic of the relationship is that if the distance is doubled, the required MAS will increase by a factor of 4. Long focal spot-receptor distances generally decrease image blur, patient exposure, and distortion; however, a significantly higher MAS is required.

AUTOMATIC EXPOSURE CONTROL

Many radiographic systems are equipped with an AEC circuit. The AEC is often referred to as the *phototimer*. The basic function of an AEC is illustrated in Figure 20-3.

The principal component of the AEC is a radiation-measuring device, or sensor, located near the receptor. A common type of sensor contains a piece of intensifying screen that converts the x-ray exposure into light. The sensor also contains a component that converts the light exposure into an electrical signal.

Within the AEC circuit, the exposure signal is compared to a reference value that has been set. When the accumulated exposure to the sensor (receptor) reaches the predetermined reference value, the x-ray tube is automatically turned off.

The reference exposure level is determined by two variables. One is the calibration of the AEC. A service engineer must adjust the basic reference level to match the sensitivity of the receptor. If either the intensifying screen or the film is

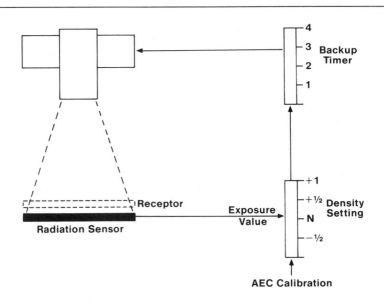

Figure 20-3 Basic automatic exposure control system.

changed, so that the overall receptor sensitivity changes, it will usually be necessary to recalibrate the AEC.

The other variable found on most systems is the *density control*, which can be adjusted by the operator. In general, the density control can be used to vary receptor exposure (film density) within a limited range. The density control is usually labeled with the factors (+ 1, − 1, etc.) by which the normal (zero setting) exposure can be changed.

The AEC contains a back-up timer, which will terminate an exposure if there is a malfunction in the normal operation. It is a safety feature to prevent excessive patient exposure. In most systems the manual timer also serves as the back-up timer for the AEC. The back-up timer should be set to a value somewhat larger than the expected exposure time. If it is set at a value that is too low, it will terminate the exposure before adequate receptor exposure has accumulated. The result will be an underexposed film.

When using AEC, the location of the sensor with respect to patient anatomy must be considered. The sensitive area of the sensor has a definite size and shape. Typically, different sensor areas can be selected by the operator. The function of the AEC is to control the average density within the sensitive area of the sensor. If the sensitive area is incorrectly positioned relative to the patient's anatomy, exposure errors can be significant. For example, in chest radiography, if the sensor field is placed over the mediastinum, the AEC will attempt to produce a density of approximately 1 unit (medium gray) in that area. Under this condition, the lung areas will be overexposed.

The use of AEC does not eliminate exposure error. Some possible sources of error that must be considered are the following:

- AEC not calibrated for a specific receptor
- density control not set to proper value
- back-up timer not set to proper value
- sensor field incorrectly positioned with respect to anatomy

---

Study Activities

Determine the MAS that will produce the same film exposure at 85 KV_p that 30 mAs produce at 100 KV_p.

Determine the MAS required to produce the same film exposure using a grid with a Bucky factor of 5 as 50 mAs produced using a grid with a Bucky factor of 2.

Determine the MAS required to produce the same film exposure as 20 mAs if the FRD is changed from 40 to 72 inches.

Explain how you would change the KV and/or MAS in going from a single-phase to a three-phase machine to produce the same film exposure.

Explain the general advantages of using a high KV as opposed to a low KV.

Explain the general reason for using low KV values for some procedures.

State when it is desirable to use a low MA value.

State when it is desirable to use high MA values.

Identify the general conditions in which the use of a short exposure time is desirable.

Identify the general conditions in which the use of a long exposure time is desirable.

Chapter 21

Fluoroscopy

The fluoroscope produces an instantaneous and continuous image that is especially useful for guiding a procedure, searching through a body section, or observing motion. Fluoroscopic examinations began soon after the discovery of x-radiation. Since that time, however, the fluoroscopic imaging system has undergone several major changes that have improved image quality, reduced patient exposure, and provided much more flexibility and ease of use.

The receptor for the first-generation fluoroscope was a flat fluorescent screen that intercepted the x-ray beam as it emerged from the patient's body. The x-ray beam carrying the invisible image was absorbed by the fluorescent material and converted into a light image. In fact, it is the fluorescent screen receptor that gives the name "fluoroscopy" to the procedure.

Under normal operating conditions, the image from the early fluoroscope had a relatively low brightness level. Because of the low light intensity, it was usually necessary to conduct examinations in a darkened room after waiting for the operator's eyes to become adjusted to the light (accomplished by wearing red goggles or remaining in the dark for approximately 20 minutes). The contrast sensitivity and visibility of detail were significantly less than what can be achieved with contemporary fluoroscopic systems.

The first major advancement was the introduction of the image intensifier tube. The intensifier tube produces a much brighter image than the fluorescent screen, and its images can be viewed in a lighted room. The quality of the image produced by the intensifier tube was generally an improvement over the fluoroscopic screen image. When the image intensifier tube was first introduced, an examination was performed by viewing the image from the intensifier tube through a system of mirrors. The viewing was generally limited to one person unless a special attachment was used.

The next step in the evolution of the fluoroscope was the introduction of a video system to transfer the image from the output of the image intensifier tube to a large screen.

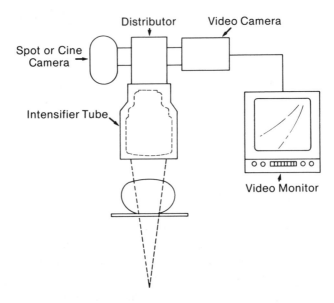

Figure 21-1 Components of a fluoroscopic receptor system.

The receptor system of a contemporary fluoroscope is shown in Figure 21-1. It consists of an image intensifier tube, an optical image distribution system, and a closed-circuit video system containing a camera, a monitor, and associated electronics. A spot film or cine camera is often included as part of the receptor system.

INTENSIFIER TUBES

We have already seen that certain fluorescent materials are used in intensifying screens to convert the x-rays into light images. Although intensifying screens are used in a wide range of radiographic applications, the brightness of the light produced by the screen is relatively low. The brightness is sufficient to expose film placed in direct contact with the screen, but the light output is generally too low for direct visualization, photographing with a camera (cine or spot film), or viewing with a television camera. In many applications a device is needed that will *convert* the x-ray into light and *intensify*, or increase the brightness of, the light in the process. The image intensifier tube is such a device.

Before considering the details of intensifier tube function, let us compare its overall function to that of a fluorescent screen, shown on the left in Figure 21-2. The tube is exposed to the x-ray beam, and light is emitted from the other end. One

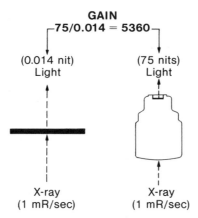

GAIN
$-75/0.014 = 5360-$

(0.014 nit)
Light

(75 nits)
Light

X-ray
(1 mR/sec)

X-ray
(1 mR/sec)

Figure 21-2 Gain characteristics of an image intensifier tube.

of the important characteristics of a specific intensifier tube is its ability to produce a bright light image.

Gain

Gain is one factor used to describe the ability of a tube to produce a bright image. As illustrated in Figure 21-2, the gain value of a specific tube is the ratio of its light brightness (in nits) to that of a fluorescent screen receiving the same x-ray exposure. Contemporary intensifier tubes have gains of 5,000 or more.

The brightness of the light from the intensifier tube is several thousand times brighter than from an old-style fluorescent screen. This is achieved in two ways.

A fluorescent screen is a passive device that converts a portion of the absorbed x-ray energy into light energy. On the other hand, an image intensifier tube is an active device that adds energy to the process. This additional energy enters the tube in the form of electrical energy from a high-voltage energy source.

Minification

The second factor that contributes to the increase in image brightness is the minification of the image as it passes through the tube. Minification is the opposite of magnification: It is when you start with a large image and reduce it to a small one. Because the same energy is concentrated in a smaller area, it becomes more intense. The light image put out by an intensifier tube appears on a small screen with a diameter of approximately 1 inch. The input field of view of the intensifier tube is much larger and generally has a diameter in the range of 4 to 14 inches. The amount of minification gain is the ratio of the areas of the input image to the output image. For example, a tube with a 5-inch-diameter input and a 1-inch-diameter output screen will produce a minification gain of 25.

Intensifier Tube Function

The method by which electrical energy is used to intensify the image is illustrated in Figure 21-3. The intensifier tube body is essentially a glass bottle with a vacuum inside. The large area forming the bottom of the bottle is the input screen, and the small area that forms the "cap" on the bottle is the output screen. These screens are also known as *phosphors*.

The input surface, or screen, of the intensifier tube is in two layers. The first layer encountered by the x-ray beam is a fluorescent material, typically cesium iodide. The second layer, which is in direct contact with the fluorescent screen, is a layer of material that functions as a photocathode. A photocathode is a material that emits electrons when it is exposed to light.

The x-ray photons entering the tube are absorbed by the fluorescent input phosphor. A portion of the absorbed energy is converted into light. Since light photons contain much less energy than x-ray photons, one x-ray photon can produce a light flash consisting of many light photons. The light photons are, in

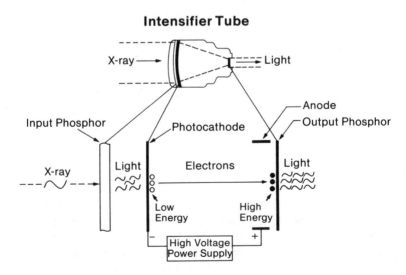

Intensifier Tube

Figure 21-3 The events that produce electronic gain in an image intensifier tube.

turn, absorbed by the photocathode layer by the photoelectric process. This causes electrons to be emitted from the cathode into the space within the tube. At this point, the electrons possess very little kinetic energy.

The intensifier tube is connected to an electrical energy source (power supply) that applies a relatively high voltage between the cathode surface and an anode located near the output end of the tube. The electrical energy accelerates the entering electrons as they travel toward the small output end of the tube, where they strike the output screen. The output screen absorbs the electron energy and converts it into a relatively bright flash of light.

In a simple fluorescent screen, the x-ray energy is converted directly into light energy. In the intensifier tube three steps are added to the process. These are

1. transferring energy from light to electrons in the photocathode
2. adding electrical energy to the electrons
3. converting electron energy back to light within the output screen

These additional steps are necessary to add energy, or intensify the image. It is not possible to increase the energy of photons. It is possible, however, to increase the energy of electrons. The result of this process is that an x-ray photon can produce a much brighter light at the output end of an intensifier tube than in a fluorescent screen.

Along the length of the tube is a series of metal bands or electrodes that focus the electron image onto the output screen. The voltage applied to the focusing electrodes can be switched to change the size of the input image, or field of view, as shown in Figure 21-4. The maximum field of view is determined by the diameter of the tube.

In most tubes, the input image area can be electrically reduced, as shown in Figure 21-4. When the tube is switched from one mode to another, the factors that change include field of view, image quality, and receptor sensitivity (exposure). The tube should be operated in the large field mode when maximum field of view is the primary consideration. In this mode the tube has the highest gain and requires the lowest exposure because the minification gain is proportional to the area of the input image. The small field mode is primarily used to give the best possible image quality.

IMAGE QUALITY

Contrast

We know that the contrast in an image delivered to the film is reduced by both object penetration and scattered radiation. When image intensifier tubes are used,

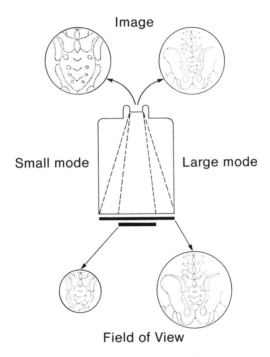

Figure 21-4 A dual-mode (FOV) image intensifier tube.

there is an additional loss of contrast because of events taking place within the tube. Some of the radiation that penetrates the input screen can be absorbed by the output screen and can produce an effect similar to scattered radiation. Also, some of the light produced in the output screen travels back to the photocathode and causes electrons to be emitted. These electrons are accelerated to the output screen where they contribute additional exposure to the image area and further reduce contrast. The contrast reduction in modern intensifier tubes is generally in the range of 5 to 15%.

Image Detail

There are several potential sources of blur and loss of detail within the system. The spreading of light in the input and output screens of the image intensifier tube produces blur, as it does in intensifying screens. With dual-mode tubes, the larger field generally produces more blur and less detail than the smaller field.

Blur is also produced by improper focusing of the electrons in the tube. Most intensifier systems have controls that can be used by the service engineer to adjust the electron focus.

Noise

The problem of quantum noise in intensified systems was discussed in Chapter 19. If it were not for the limitations of quantum noise, modern intensifier systems could operate at much lower input exposure values.

VIDEO SYSTEMS

The primary function of a video system is to transfer an image from one location to another. During the transfer process, certain image characteristics, such as size, brightness, and contrast, can be changed. However, as an image passes through a video system, there can be loss of quality, especially in the form of blur and loss of detail visibility.

Video Principles

The two major components of a video system are the camera and the monitor, or receiver. Conventional broadcast television systems transmit the image from the camera to the receiver by means of radio frequency (RF) radiation. In a closed circuit system, the image is transmitted between the two devices by means of electrical conductors or cables. However, other than the means of image transmission, the basic principles of the two systems are essentially the same.

A basic video system is illustrated in Figure 21-5, which shows the major functional components of the camera and the monitor. The "heart" of each is an electronic tube that converts the image into an electrical signal or vice versa.

The function of the camera tube is to convert the light image into an electronic signal.

The typical camera tube is cylindrical, with a diameter of approximately 25 mm and a length of 15 cm. The image to be transmitted is projected onto the input screen of the tube by a lens like that in a conventional film camera. The other end of the tube contains a heated cathode and other electrodes that form an *electron gun*. The electron gun shoots a small beam of electrons down the length of the evacuated tube. The electron beam is moved across the screen surface on which the input image is projected.

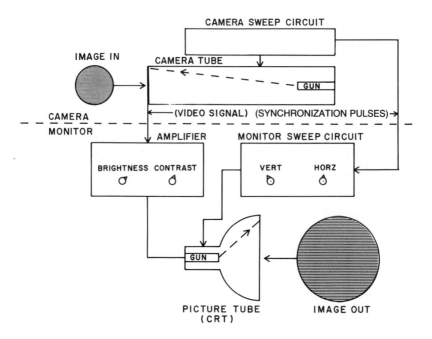

Figure 21-5 Basic components and function of a fluoroscopic video system.

Image Scanning

The beam begins in the upper left-hand corner and moves across the image horizontally, as shown in Figure 21-6. In the conventional American 525-line video system, lines are scanned at the rate of 15,750 per second. When the beam reaches the right-hand side, it is quickly "snapped back" to the left and deflected downward by approximately one beam width. It then sweeps across to form a second scan line. This process is repeated until the beam reaches the bottom of the screen, which usually requires 1/60th of a second. After reaching the bottom, the beam returns to the top and resumes the scanning process. This time, however, the scan lines are slightly displaced with respect to the first set so that they fall between the lines created in the first scan field. This is known as *interlacing*.

Interlacing is used to prevent *flicker* in the picture. If all lines were scanned consecutively, it would take twice as long (1/30th of a second) for the beam to reach the bottom of the screen. This delay would be detectable by the human eye and would appear as flicker. With interlacing, the face of the screen is scanned in two sets of lines, or fields. The pattern of scan lines produced in a video system is known as the *raster*. The conventional video image is generally said to contain 525

INTERLACED SCAN

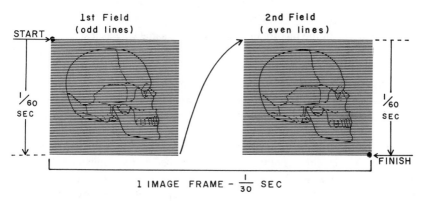

Figure 21-6 An interlaced scan format.

lines; however it generally contains only 485 lines. This is because of the time lost during the return of the electron beam to the left side of the screen. In a 525-line system, 30 complete raster frames (60 fields) are formed per second.

Video Signal

The screen of the camera tube is made of a material with light-sensitive electrical properties. Several types of tubes are used, but the general concept of tube function can be described as follows.

The electrical conductivity of the screen surface depends on its illumination. When an image is projected onto the screen, the conductivity varies from point to point. A dark area is essentially nonconductive, and a brightly illuminated area is the most conductive. As the electron beam sweeps over the surface, it encounters areas with various levels of conductivity, which depend on the brightness at each point. When the beam strikes a bright spot, it is "conducted through" and creates a relatively high signal voltage at the output terminal of the tube. As the spot moves across the screen, it creates a signal that represents the brightness of the image at each point along its path.

Picture Tube Function

The heart of the receiver, or monitor, is the picture tube. The picture tube differs from the camera tube in size and shape. One end of the tube is the screen on which the video image is displayed. Like the camera tube, the picture tube has an electron gun located in the end opposite to the screen. The electron gun produces a beam of electrons that strikes the rear of the screen in the picture tube. The electron beam scans the surface of the picture tube screen in the same way as it scans the camera tube screen. In fact, the scanning in the two tubes is synchronized by a signal transmitted from the camera to the monitor. If the scanning becomes unsynchronized, the image will roll in the vertical direction or become distorted horizontally. The horizontal and vertical controls on a video monitor are used to adjust the scan rates so that they are identical with those of the camera and can maintain synchronization.

When the electron beam in the picture tube strikes the screen, it produces a bright spot. The brightness of the spot is determined by the number of electrons in the beam, which is controlled by the signal from the camera tube. In other words, the brightness of a spot on the picture tube screen is determined by the brightness of the corresponding point on the camera tube screen. As the two electron beams scan the two screens, the image is transferred from the camera tube to the picture tube. In the 525-line system, complete images are transferred at the rate of 30 per second.

Contrast

In the typical video system, image contrast can be changed by adjusting a control located in the monitor. The average video signal level is changed by adjusting the brightness control. Although the contrast and brightness controls are essentially separate and independent, they must generally be adjusted together for optimum image quality.

Blur and Visibility of Detail

One generally undesirable characteristic of a video imaging system is that it introduces blur into the image and reduces detail. One source produces blur in the vertical direction, and another source produces it in the horizontal. Although the blur values can be different for the two directions, the overall image quality is usually best when they are approximately equal.

Vertical Blur

Vertical blur is caused by the size of the electron beam and the width of the scan lines. The effect of vertical blur is illustrated in Figure 21-7. If a small-line-type object is oriented at a slight angle to the scan lines, the images of the object will

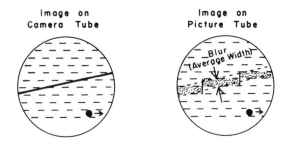

Figure 21-7 Vertical blur produced by scan lines.

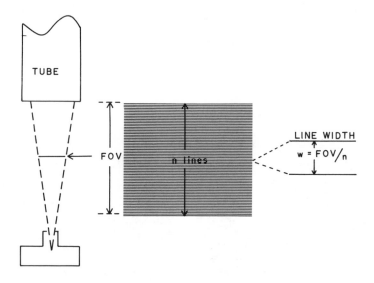

Figure 21-8 The factors that affect line width and vertical detail in a video image.

appear to be wider because of blur. At any instant, the width of an object in the image cannot be less than the width of one scan line. For an image containing a given number of scan lines, vertical blur is directly proportional to the dimension of the image, or the field of view, as illustrated in Figure 21-8.

Special attention is called to the fact that video blur is directly proportional to the field of view at the input to the image intensifier tube, as illustrated in Figure 21-9. A small field of view produces better detail because the lines are smaller.

Blur can be decreased by increasing the number of lines used to form the video image. Figure 21-10 illustrates two of the most common video scan patterns used

SMALL FIELD OF VIEW   LARGE FIELD OF VIEW

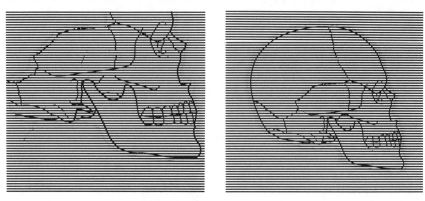

Figure 21-9 Improvement in image detail using a small field of view.

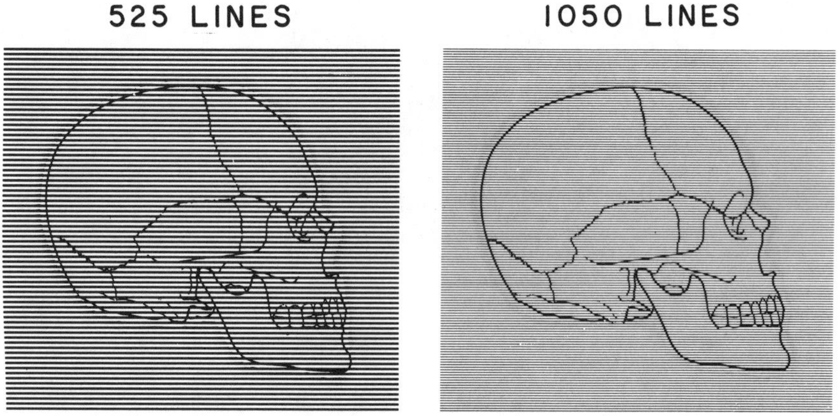

525 LINES   1050 LINES

Figure 21-10 Improvement in image detail by increasing the number of scan lines.

in fluoroscopy. Most are classified as 525-line systems. In special applications that require more image detail, 1,050-line systems are used.

Horizontal Blur

Blur and detail in the horizontal direction are determined by the response time of the electronic circuitry through which the video signal passes in going from the camera to the picture tube. In most systems the horizontal blur is adjusted to be approximately equal to the vertical blur. This cannot be adjusted by the operator.

Noise

The two types of noise in a fluoroscopic system are electronic and quantum. Electronic noise produces "snow," which is familiar to most television viewers. It is usually significant only when the video signals are extremely weak. Since signal strength is not a problem in the typical closed circuit video system, the presence of significant "snow" or electronic noise usually indicates problems within the video system.

Quantum noise depends on the number of photons used to form the image. The number of photons involved in image formation is directly related to the receptor input exposure. The input exposure, for a specific image brightness, can be adjusted by changing the automatic brightness control circuit reference level, as discussed in Chapter 19. An input exposure rate of approximately 0.025 mR/sec (1.5 mR/min) is usually required to reduce quantum noise to an acceptable level in fluoroscopy.

The noise level is related to the total number of photons used to form an image, not the rate. The human eye has an effective "collection" time of approximately 0.2 seconds. This means we do not see each video image frame as it is flashed on the screen. We see several frames added together by our visual system. Because the several frames represent more photons than a single frame, we see less quantum noise.

THE OPTICAL SYSTEM AND CAMERAS

An optical system is used to transfer the image from the output screen of the intensifier tube to the input screen of the video camera tube or to the film in the spot or cine camera. The components of the total optical system are contained in the image distributor and the individual cameras, as shown in Figure 21-11, and are lenses, apertures, and mirrors. Before we consider the operation of the optical system, we will review the basic characteristics of two of these components, lenses and apertures.

Lens

A lens is the basic element that can transfer an image from one location to another. The curvature of the lens focuses the light that passes through it.

A fundamental characteristic of a lens is its *focal length*, which expresses its focusing power. The focal length is the distance between the lens and the point at which all parallel light rays that enter the lens are brought together, or focused. The focal length is determined by the curvature of the lens and is typically expressed in millimeters.

The focal length of a lens is a major factor in determining the size of an image projected onto a film or screen.

Aperture

Another important characteristic of a lens is its size (diameter, or aperture). This determines the amount of light that is captured by the lens. This in turn affects the efficiency of light transfer through the optical system and the exposure to film in the camera.

In many applications it is desirable to adjust the aperture size, or light-gathering efficiency, of a lens. This is accomplished by covering a portion of the lens with a circular mask with a hole in the center. In conventional photographic cameras, the aperture can usually be adjusted to different sizes. In many radiographic systems, the size of the aperture is changed by replacing one mask with another.

Image Distributor

When intensifying screens are used in radiography, the light is transferred directly to the film because the film is in direct contact with the screen. This results in a film image that is the same size as the image from the intensifying screen. The output image from the typical intensifying tube, however, is only about 20 mm in diameter and must be enlarged before it is applied to the film. In order to be enlarged, the film must be separated from the output screen. The transfer of the image from the image intensifier output to the film surface is the main function of the optical system. If only one camera is to be used, it is a relatively simple process to mount the camera so that it views the image from the intensifier tube. With many fluoroscopic systems, however, it is desirable to transfer the image to spot film or cine cameras in addition to the video camera. This is the function of the part of the optical system known as the image distributor.

Collimator Lens

The first component of the optical system encountered by the light from the intensifier output screen is the collimator lens. Its function is to collect the light from the output screen and focus it into a beam of parallel light rays, as shown in Figure 21-11. The formation of the image into a parallel beam makes it possible to distribute the image to two or more devices, such as a spot film camera and a video camera, for fluoroscopy.

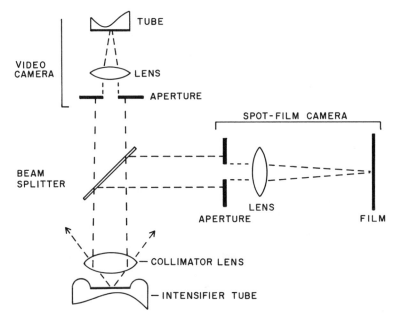

Figure 21-11 Basic optical system for a fluoroscope.

Mirrors

The next element in the path of the light is a beam splitter. A splitter is usually a mirror that is only partially reflective. A portion of the light is reflected by the mirror to one camera. The remaining light passes through the mirror to a second camera. In some systems the mirror is attached to a rotating mechanism so that it can be shifted from one camera to another. The mirrors are generally designed to divide the light unevenly between two devices. For example, a 70-30 mirror sends 70% of the light to a film camera and 30% to a video camera to form the fluoroscopic image.

Camera

After passing through the aperture, a second lens focuses the light onto the surface of the film to form the final image. This lens is part of the camera and serves the same function as a conventional camera lens. The most significant characteristic of this lens is its focal length.

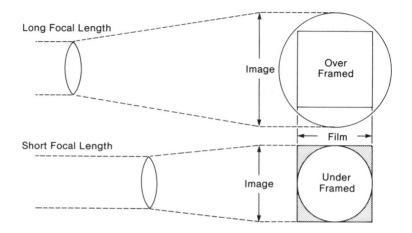

Figure 21-12 Relationship of image and film size for different degrees of framing.

Image Size and Framing

The amount of enlargement, or magnification, in the image between the image intensifier output and the film is given by

Magnification (m) = Focal length of camera lens/Focal length of collimator lens

The camera lens must be selected to give the desired image size on the film, as shown in Figure 21-12. Four film sizes are generally used in intensified radiography: 35 mm, 70 mm, 90 mm, and 105 mm. These are the dimensions of the film. The size of the image area is somewhat less because of sprocket holes and borders. For example, 35-mm film typically has an image area of 25 by 18 mm. The image at the intensifier output screen is circular, but all films have either square or rectangular image areas. Because of this, it is impossible to get exact coverage on the film.

For a given film size, it is possible to obtain different degrees of coverage, or framing. With underframing, all of the image appears on the film but is circular; the corners of the film are not exposed. On the other hand, with overframing, all of the film is used, but some portions of the image are lost.

Unless the optical system is properly focused, it will also be a source of image blur. Both the collimator and camera lens can be out of focus. Proper focusing of the collimator lens requires special equipment and should be attempted only by qualified personnel.

RECEPTOR SENSITIVITY

Fluoroscopy

The sensitivity during fluoroscopic operations is set by the equipment engineer through an adjustment of the video camera sensitivity or the video camera aperture. The sensitivity and exposure also change with the field of view (mode), as described above. The fluoroscope is most sensitive when operated with the maximum field of view. Increasing field of view increases sensitivity and decreases required exposure. Because the x-ray beam then covers more of the patient's body, however, the total radiation energy to the patient is not significantly reduced.

Some fluoroscopic systems have a control that allows the operator to change the sensitivity. This is used to control the level of quantum noise in the fluoroscopic image. The low sensitivity (low noise) settings are used to improve visibility in certain demanding procedures, such as angioplasty.

Radiography

The overall sensitivity of an intensified radiography receptor system depends on three major factors: (1) the gain or conversion factor of the image tube, (2) the efficiency of the optical system, and (3) the sensitivity of the film.

Within a given system, the receptor sensitivity can usually be changed by adjusting the aperture. If the film is changed to one with a different sensitivity, the aperture can be used to compensate for the change and to maintain the same receptor sensitivity, if desired.

Study Activities

Name two functions performed by an image intensifier tube.

Explain how an image intensifier tube increases the brightness (intensity) of an image.

Explain how the gain of an image intensifier tube is affected by the field of view.

Explain the process that a video system uses to transfer an image from one place to another.

Identify the image quality characteristic affected by the width of the video scan line.

Explain how changing the field of view in fluoroscopy affects image detail.

Discuss the reason for having a collimator lens in a fluoroscopy system.

Identify the component within the optical system that controls the amount of light reaching the film (exposure).

Identify the component within the optical system that affects the size of the image projected onto the film.

Explain how you could identify an overframed or underframed image.

Chapter 22

Tube Heating and Cooling

To produce x-radiation, relatively large amounts of electrical energy must be transferred to the x-ray tube. Only a small fraction (typically less than 1%) of the energy deposited in the x-ray tube is converted into x-rays; most appears in the form of heat. This places a limitation on the use of an x-ray apparatus. If excessive heat is produced in the x-ray tube, the temperature will rise above critical values, and the tube can be damaged. This damage can be in the form of a melted anode or a ruptured tube housing. In order to prevent this damage, the x-ray equipment operator must be aware of the quantity of heat being produced and its relationship to the *heat capacity* of the x-ray tube.

Figure 22-1 identifies the factors that affect both heat production and heat capacity.

HEAT PRODUCTION

Heat is produced in the focal spot area by the bombarding electrons from the cathode. When the electrons arrive at the anode surface they are carrying kinetic energy. When they hit the anode and come to a sudden stop their kinetic energy must be transferred to other forms. Most of it is converted to heat. A small amount, usually not more than 1% of the electrons' energy, is converted to x-radiation. The amount of heat that is produced in the focal spot can be determined by calculating the total electrical energy transferred during the exposure period. We can neglect the small amount of energy that goes into x-radiation.

Heat Units

There are two units used to express the quantity of heat in an x-ray tube. One is the joule (J). The joule is the basic unit for energy; this is appropriate because heat

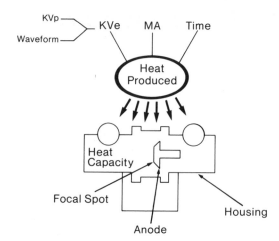

Figure 22-1 Factors that determine the amount of heat produced and the three areas of an x-ray tube that have specific heat capacities.

is one form of energy. The other unit is the heat unit (HU), a special unit that was introduced specifically to express x-ray tube heat. These two units are not the same size. One joule represents more heat than one heat unit. Because the heat unit is smaller the number expressing a specific quantity of heat in heat units would be larger than the number expressing the same amount of heat in joules. If you know an amount of heat in one unit you can calculate the amount in the other unit by:

$$\text{heat (HU)} = 1.4 \times \text{heat (J)}$$
$$\text{or}$$
$$\text{heat (J)} = 0.71 \times \text{heat (HU)}$$

Example:

$$1,000 \text{ J} \times 1.4 = 1,400 \text{ HU}$$
$$\text{or}$$
$$1,000 \text{ HU} \times 0.71 = 710 \text{ J}$$

The amount of heat produced during each exposure is proportional to three factors: (1) the KV, (2) the MA, and (3) the exposure time. Therefore, we control the amount of heat produced by these three technique factors.

KV

Let us recall that the KV determines the amount of energy carried by each electron to the x-ray tube. Since the heat is produced by the electrons' energy, it is also proportional to the KV. The amount of heat produced is most directly related to the effective KV (KV$_e$) rather than the peak KV (KV$_p$). You will recall that the relationship between effective and peak KV depends on the electrical waveform, such as single-phase or three-phase. Three-phase electricity produces a higher effective KV than does single-phase because it is more constant throughout the exposure. The effective KV is always some fraction of the peak KV. This fraction is known as the *waveform factor*. Values for the waveform factors are

- constant potential   1
- three-phase, 12-pulse   0.99
- three-phase, 6-pulse   0.96
- single-phase   0.71

The significance of this is that a certain KV$_p$ value will produce more heat if it is three-phase electricity than if it is single-phase.
Example:

$$100 \text{ KV}_p \text{ three-phase} = 99\text{KV}_e$$
$$100 \text{ KV}_p \text{ single-phase} = 71 \text{ KV}_e$$

We have just seen that the amount of heat produced is proportional to the KV and is greater for three-phase electricity than for single-phase. However, we must be careful not to draw an incorrect conclusion from this. In actual practice the heat produced is less when we use a high KV technique rather than a low one. It is also less when we use three-phase rather than single-phase equipment. The reason for this is that using either a high KV or three-phase equipment allows us to use a lower MAS, thus reducing the total heat. We should recall that the efficiency of x-ray production increases as we increase KV and is also higher for three-phase than single-phase electricity. This means that more of the electron energy is going into the production of radiation rather than the production of heat. This radiation is also more penetrating and less is needed to expose a film.

MAS

The heat produced during a single exposure is proportional to both MA and exposure time. Recall that these two factors determine the total quantity of electrons that flow through the tube during an exposure. Since each electron brings some energy that is converted into heat, total heat produced is proportional to MAS.

Calculations

We can calculate the amount of heat produced in each exposure by multiplying three factors: (1) KV_p, (2) waveform factor, and (3) MAS, as shown below.

$$\text{heat (J)} = KV_p \times \text{waveform factor} \times \text{MAS}$$

Examples

heat (J)	=	$100\ KV_p \times 0.96 \times 50\ \text{MAS}$	Three-phase, 6-pulse
heat	=	4,800 J	
heat (J)	=	$100\ KV_p \times 0.71 \times 50\ \text{MAS}$	Single-phase
heat	=	3,550 J	

You recall that heat measured in joules can be converted into heat units by multiplying by 1.4. If we multiply the waveform factor for single-phase electricity (0.71) by 1.4, we get a value of 1 (0.71 × 1.4 = 1). This makes it possible to calculate the heat produced in single-phase equipment in heat units by using the simple formula

$$\text{Heat (HU)} = KV_p \times \text{MAS}$$

In fact, the heat unit was invented specifically to simplify this calculation.

Heat Capacity

In order to evaluate the problem of x-ray tube heating, it is necessary to understand the relationship of three physical quantities: (1) heat, (2) temperature, and (3) heat capacity. Heat is a form of energy and can be expressed in any energy units. In x-ray equipment, heat is usually expressed in joules (watt-seconds) or heat units.

Temperature is the physical quantity associated with an object that indicates its relative heat content. Temperature is specified in units of degrees. Physical changes, such as melting, boiling, and evaporation, are directly related to an object's temperature rather than its actual heat content.

For a given object, the relationship between temperature and heat content involves a third quantity, heat capacity, which is a physical characteristic of the object. The general relationship can be expressed as follows:

$$\text{Temperature} = \text{heat/heat capacity}$$

The heat capacity of an object is more or less proportional to its size (or mass) and a characteristic of any given material known as its *specific heat*. As heat is added to

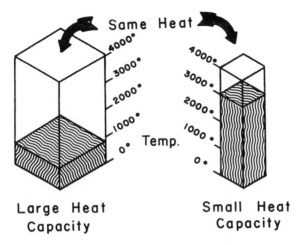

Figure 22-2 Relationship among heat, temperature, and heat capacity.

an object, the temperature increases in proportion to the amount of heat added. When a given amount of heat is added, the temperature increase is *inversely* proportional to the object's heat capacity. In an object with a large heat capacity, the temperature rise is smaller than in one with a small heat capacity. In other words, the temperature of an object is determined by the relationship between its *heat content* and its *heat capacity*. This is illustrated in Figure 22-2.

In x-ray tube operation the goal is never to exceed specific critical temperatures that produce damage. This is achieved by keeping the heat content below specified critical values related to the tube's heat capacity.

In most x-ray tubes there are three distinct areas with critical heat capacities, as shown in Figure 22-3. The area with the smallest capacity is the focal spot area, or *track*, and it is the point at which heat is produced within the tube. From this area, the heat moves by *conduction* throughout the anode body and by *radiation* to the tube housing; heat is also transferred by radiation from the anode body to the tube housing. Heat is removed from the tube housing by transfer to the surrounding atmosphere. When the tube is in operation, heat generally flows into and out of the three areas shown. Damage can occur if the heat content of any area exceeds its maximum heat capacity.

FOCAL SPOT AREA

The maximum heat capacity of the focal spot area, or track, is the major limiting factor with single exposures. If the quantity of heat delivered during an individual exposure exceeds the track capacity, the anode surface can melt, as shown in

HEAT   CAPACITY   AND   TRANSFER

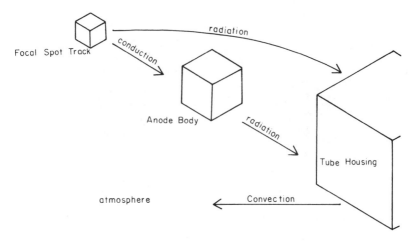

Figure 22-3 The three critical heat capacities in an x-ray tube.

Figure 22-4. The capacity of a given focal spot track is generally specified by the manufacturer in the form of curves, as shown in Figure 22-5. This type of curve shows the maximum power (KV and MA) that can be delivered to the tube for a given exposure time without producing overload. Graphs of this type are generally designated *tube rating charts*. From the graph it is seen that the safe KV and MA limit of a tube is inversely related to the exposure time. This is not surprising since the total heat developed during an exposure is the product of KV, MA, and exposure time. It is not only the total amount of heat delivered to the tube that is crucial but also the time in which it is delivered.

An x-ray tube will usually have several different rating charts, because a tube can be operated under different conditions that affect heat capacity. The three factors affecting heat capacity are (1) focal spot size, (2) anode rotation speed, and (3) the kilovoltage waveform. Each of these is discussed in more detail below. When using a rating chart, you must know how the tube is being operated with respect to these three factors so that you can select the appropriate chart.

X-ray tubes are often given single power ratings in kilowatts (kW). By general agreement an exposure time of 0.1 second is used for specifying a tube's power rating, the product of KV and MA. Although this does not describe a tube's limitations at other exposure times, it does provide a means of comparing tubes and operating conditions.

A number of different factors determine the heat capacity of the focal spot track. The focal spot track is the surface area of the anode that is bombarded by the electron beam. In stationary anode tubes it is a small area with dimensions of a few

Figure 22-4 Rotating anode damaged by overheating.

Single Exposure Ratings
Single-Phase Full-Wave Rectification

0.3 mm Focal Spot
High Speed Rotation

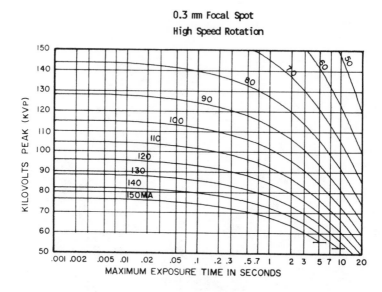

0.3 mm Focal Spot
Low Speed Rotation

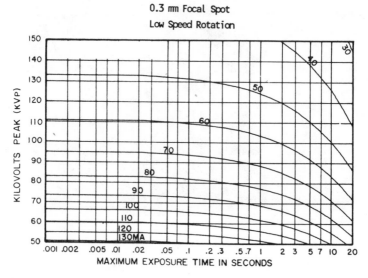

Figure 22-5 Single exposure rating chart. *Source*: Courtesy of Varian-Eimac, Salt Lake City, UT.

Figure 22-5 continued

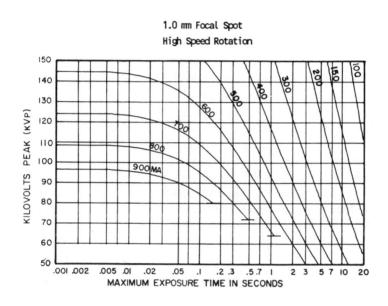

1.0 mm Focal Spot
High Speed Rotation

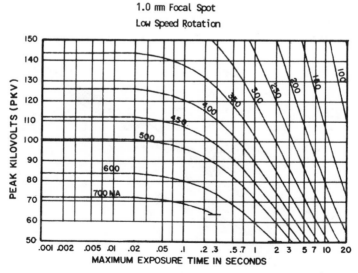

1.0 mm Focal Spot
Low Speed Rotation

millimeters. In the rotating anode tube the focal spot track is much larger because of the movement of the anode with respect to the electron beam. Figure 22-6 shows a small section from a rotating anode disk. The surface is at an angle with respect to the central ray of the beam.

Focal Spot Size

From the standpoint of producing x-ray images with minimum blur, a small focal spot is desired. However, a small focal spot tends to concentrate heat and give the focal spot track a lower heat capacity. The only advantage of a larger focal spot is increased heat capacity. Many x-ray tubes have two focal spot sizes that can be selected by the operator. The small focal spot is generally used at relatively low power (KV and MA) settings. The large focal spot is used when the machine must be operated at power levels that exceed the rated capacity of the small focal spot. The specified size of an x-ray tube focal spot is the dimensions of projected focal spot shown in Figure 22-6. This is the area you would see if you were to look from the receptor. Notice that the actual focal spot, the area bombarded by the electron beam, is always larger than the projected, or effective, focal spot. For a given anode angle, the width of the focal spot track is directly proportional to the size of

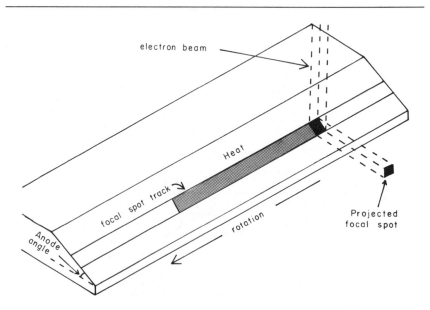

Figure 22-6 Section of a rotating anode showing relationship of focal spot track to electron beam and anode angle.

the projected spot. The relationship between heat capacity and specified focal spot size is somewhat different. In many tubes, doubling the focal spot size increases the power rating by a factor of about 3.

Anode Angle

The actual relationship between focal spot width (and heat capacity) and the size of the projected focal spot is determined by the anode angle. Anode angles generally range from about 7 to 20 degrees. For a given effective focal spot size, the track width and heat capacity are inversely related to anode angle. Although anodes with small angles give maximum heat capacity, they have specific limitations with respect to the area that can be covered by the x-ray beam. X-ray intensity usually drops off significantly toward the anode end because of the *heel effect*. In tubes with small angles, this is more pronounced and limits the size of the useful beam. Figure 22-7 shows the nominal field coverage for two different anode angles. The x-ray tube anode angle should be selected by a compromise between heat capacity and field of coverage.

Anode Rotation Speed

In rotating tubes, the anode assembly is mounted on bearings and actually forms the *rotor* of an electric motor. The x-ray tube is surrounded by a set of coils that

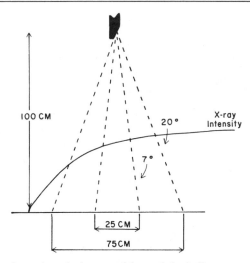

Figure 22-7 Variation of x-ray intensity because of the anode heel effect.

forms the *stator* or nonmoving part of the motor. When the coils are energized from the power line, the motor spins. The speed of rotation is determined by the frequency of the applied current. When the stator coils are operated from the 60-Hz power line, the speed of rotation is approximately 3,000 rpm. By using a special power supply that produces 180-Hz current, rotation speeds of approximately 10,000 rpm can be obtained. This is commonly referred to as high-speed rotation.

The *effective* length of the focal spot track in which the heat is deposited is proportional to the speed of rotation for a given exposure time. High-speed rotation simply spreads the heat over a longer track, especially in short exposure times. High-speed rotation generally increases the power capacity of a tube by approximately 60%.

Kilovoltage Waveform

Another factor that affects the heat capacity of the focal spot track is the waveform of the KV. Single-phase power delivers energy to the anode in pulses, as shown in Figure 22-8. Three-phase power delivers the heat at an essentially constant rate. Figure 22-8 compares the temperatures produced by a single-phase and a three-phase machine delivering the same total heat. Because of the pulsating nature of single-phase power, some points on the anode surface are raised to higher temperatures than others. These hot spots exceed the temperature produced by an equal amount of three-phase energy. When an x-ray tube is operated from a single-phase power supply, the maximum power must be less than for three-phase operation, to keep the hot spots from exceeding the critical temperature. In other words, three-phase operation *increases* the effective focal spot track heat capacity and the rating of an x-ray tube.

The effect of kilovoltage waveform on tube rating should not be confused with the effect of waveform on heat production, which was discussed earlier. However, both factors should be considered to determine if there is any advantage, from the

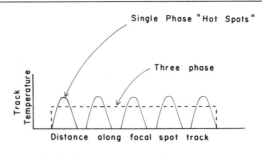

Figure 22-8 Approximate distribution of temperature along the focal spot track for single-phase operation.

standpoint of tube heating, in using three-phase power. In comparing three-phase and single-phase operation, three factors should be considered:

1. Three-phase operation permits a tube to be operated at a higher power level because of the uniform distribution of heat.
2. Three-phase operation produces more x-radiation and increased penetration at a given KV_p and MAS setting.
3. Three-phase operation produces more heat for a given KV_p and MAS setting.

The real advantage of three-phase operation is related to the first two factors. Because of the increased efficiency of x-ray production and the increased penetrating ability of the radiation, a lower KV_p or MAS value is required to produce a given film exposure. This more than compensates for the increased waveform factor and heat production associated with three-phase operation. The increased rating, or maximum permissible power, associated with the three-phase waveform also adds to the advantage. An x-ray tube can generally be operated at a higher power level when the power is supplied from a three-phase power supply, and it will also produce radiation more efficiently.

A rating chart for an x-ray tube operated at different waveforms and rotation speeds is shown in Figure 22-5. The highest power capacity is obtained by using three-phase power and high-speed rotation; notice that the real advantage occurs at relatively short exposure times. As exposure time is increased, overlapping of the focal spot track and the diffusion of heat make the difference in power capacity much less significant.

Most rotating anode tubes contain two focal spots. As mentioned previously, the size of the focal spot significantly affects the heat capacity. Remember that a given x-ray tube focal spot track has a number of different heat capacities or rating values, depending on focal spot size, rotation speed, and waveforms. Some typical values are shown in Table 22-1. The values in the table indicate the

---

Table 22-1 Heat Rating (in Joules) for Typical X-Ray Tube for Exposure Time of 0.1 sec and Focal Spot Sizes of 0.7 mm and 1.5 mm

	Single-phase	Three-phase
3,600 rpm	700 (0.7 mm)	1,050 (0.7 mm)
	2,300 (1.5 mm)	3,400 (1.5 mm)
10,800 rpm	1,100 (0.7 mm)	1,700 (0.7 mm)
	3.900 (1.5 mm)	5,800 (1.5 mm)

Source: Reprinted from *Physical Principles of Medical Imaging* (p 137) by P Sprawls Jr, Aspen Publishers, Inc, © 1987.

maximum safe heat that can be produced in the same x-ray tube for eight different operating conditions. Notice that the lowest heat capacity occurs with the small focal spot, normal speed rotation (3,600 rpm), and single-phase operation. The heat capacity for this particular tube can be increased by more than eight times when using a large focal spot, three-phase operation, and high-speed rotation.

ANODE BODY

The heat capacity of the focal spot track is generally the limiting factor for single exposures. In a series of radiographic exposures, computed tomography (CT) scanning, or fluoroscopy, the build-up of heat in the anode body can become significant. Excessive anode temperature can crack or warp the anode disc. The heat capacity of an anode is generally described graphically, as shown in Figure 22-9. The set of curves, describing the thermal characteristics of an anode, conveys several important pieces of information. The maximum heat capacity is indicated on the heat scale. The heating curves indicate the build-up of heat within the anode for various energy input rates. These curves apply primarily to the continuous operation of a tube, such as in CT or fluoroscopy. For a given x-ray tube there is a critical input rate that can cause the rated heat capacity to be exceeded after a period of time. This is generally indicated on the graph. If the heat input rate is less than this critical value, normal cooling prevents the total heat content from reaching the rated capacity.

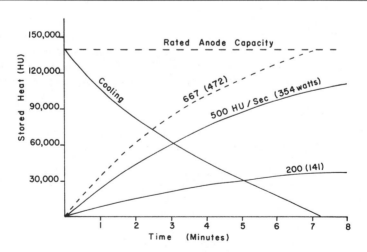

Figure 22-9 Anode heating and cooling curves.

The cooling curve can be used to estimate the cooling time necessary between sets of exposures. Suppose a rapid sequence of exposures has produced a heat input of 90,000 HU. This is well over 50% of the anode storage capacity. Before a similar sequence of exposures can be made, the anode must cool to a level at which the added heat will not exceed the maximum capacity. For example, after an initial heat input of 90,000 HU, a cooling time of approximately 3.5 minutes will decrease the heat content to 30,000 HU. At this point another set of exposures producing 90,000 HU could be taken.

The cooling rate is not constant. An anode cools faster when it has a high heat content and a high temperature. In CT scanning, when anode heating is a limiting factor, a higher scan rate can be obtained by operating the anode with the highest safe heat content, since the cooling rate is higher for a hot anode and more scans can be obtained in a specific time than with a cool anode. Most CT systems have a display that shows the anode heat content as a percentage of the rated capacity.

The anodes in most radiographic equipment are cooled by the natural radiation of heat to the surrounding tube enclosures. However, anodes in some high-powered equipment, such as that used in CT, are cooled by the circulation of oil through the anode to a heat exchanger (radiator).

Anode damage can occur if a high-powered exposure is produced on a cold anode. It is generally recommended that tubes be warmed up by a series of low-energy exposures to prevent this type of damage.

Tube Housing

The third heat capacity that must be considered is that of the tube housing. Excessive heat in the housing can rupture the oil seals, or plugs, and cause hot oil to run out. Like the anode, the housing capacity places a limitation on the extended use of the x-ray tube rather than on individual exposures. Since the housing is generally cooled by the movement of air (convection), its effective capacity can be increased by using forced-air circulation.

The housing heat capacity is much larger than that of the anode and is typically over 1 million HU. The time required for a housing to dissipate a given quantity of heat can be determined with cooling charts supplied by the manufacturer.

SUMMARY

The heat characteristics of x-ray tubes should be considered when tubes are selected for specific applications and should be used as a guide to proper tube operation.

Study Activities

Convert 6,000 J of heat into HU.

Convert 6,000 HU of heat into joules (J).

Calculate the amount of heat that would be produced by a single-phase machine for one exposure using 90 KV_p, 200 mA, and 0.1 s.

Repeat the above calculation for a three-phase machine.

Name the three parts of an x-ray tube that have specific heat capacities.

Explain why it is desirable for a focal spot to have a large heat capacity.

Describe the possible effects of exceeding the heat capacity of a focal spot area.

Explain the basic advantage in using large focal spots.

Describe the relationship between anode rotation speed and focal-spot heat capacity.

Explain how changing from single-phase to three-phase operation affects the heat capacity of a focal spot track.

Use the rating chart in Figure 22-5 to find the maximum safe exposure time for the following conditions:

Focal Spot	Rotation	KV_p	MA
0.3	high	90	130
1.0	low	90	500
0.3	low	90	70

Using the cooling curve in Figure 22-9, determine the time required for an anode to cool from 120,000 HU to 60,000 HU.

Chapter 23

Patient Exposure and Protection

All medical imaging methods deposit some form of energy in the patient's body. Although the quantity of energy is relatively low, it is a factor that should be given attention when conducting diagnostic examinations. In general, there is more concern for the energy deposited by the ionizing radiations, x-ray and gamma, than for ultrasound energy or radiofrequency (RF) energy deposited by magnetic resonance imaging (MRI) examinations. Therefore, this chapter gives major emphasis to the issues relating to the exposure of patients to ionizing radiation.

Patients undergoing x-ray examinations are subject to a wide range of exposure levels. One of our objectives is to explore the factors that affect patient exposure. This is followed by an example of exposure values in the clinical setting.

Figure 23-1 identifies the major factors that affect patient exposure during a radiographic procedure. Some factors, such as thickness and density, are determined by the patient. Most of the others are determined by the radiographers and medical staff. Many of the factors that affect patient exposure also affect image quality. In most instances, when exposure can be decreased by changing a specific factor, image quality is also decreased. Therefore, the objective in setting up most x-ray procedures is to select factors that provide an appropriate compromise between patient exposure and image quality.

X-RAY EXPOSURE PATTERNS

In any x-ray examination, there is considerable variation in exposure from point to point within the patient's body. This must be considered when expressing values for patient exposure. In fact, when exposure values are given, the specific anatomical location of the value should also be stated. Some exposure patterns are characteristic of the different x-ray imaging methods. A review of these patterns

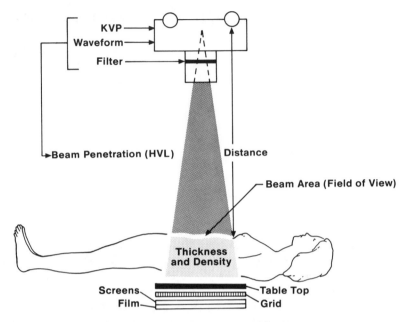

Figure 23-1 Factors that affect patient exposure in a radiographic procedure.

will give us some background for considering factors that affect exposure and applying methods to determine actual exposure values.

Radiography

In the typical radiographic examination, the x-ray beam is projected through the patient's body, as shown in Figure 23-2. The point that receives maximum exposure is the entrance surface near the center of the beam. There are two reasons for this: The primary x-ray beam has not been attenuated (absorbed) by the tissue at this point, and the area is exposed by some of the backscattered radiation from the body. The amount of surface exposure produced by the backscatter depends on the spectrum of the primary beam and the size of the exposed area. For typical radiographic situations, scattered radiation can add at least 20% to the surface exposure produced by the primary beam.

As the x-ray beam progresses through the body, it undergoes attenuation. The rate of attenuation is determined by the photon-energy spectrum (KV and filtration) and the type of tissue (fat, muscle, bone) through which the beam passes. For

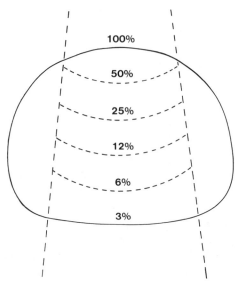

Figure 23-2 Typical exposure pattern (depth dose curves) for an x-ray beam passing through a patient's body.

the purpose of the discussion we assume a body consisting of homogeneous muscle tissue. In Figure 23-2, lines are drawn to divide the body into HVL's. The exposure is reduced by a factor of one half each time it passes through 1 HVL. The thickness of 1 HVL depends on the photon energy spectrum. However, for the immediate discussion, we assume that 1 HVL is equivalent to 4 cm of tissue. A 20-cm thick body section consists of 5 HVLs. Therefore, the exposure decreases by one half as it passes through each 4 cm of tissue. At the exit surface, the exposure is a small fraction of the entrance surface exposure.

The exposure to a specific organ or point of interest within the direct x-ray beam depends on its nearness to the entrance surface.

Tissue located outside the primary beam receives some exposure from the scattered radiation produced within the beam area. The scatter exposure to the surrounding tissue is relatively low in comparison to the exposure levels within the primary beam.

Fluoroscopy

The fluoroscopic beam projected through the body will produce a pattern similar to a radiographic beam if the beam remains fixed in one position. If the beam is moved during the procedure, the radiation will be distributed over a large volume of tissue rather than being concentrated in one area. For a specific exposure time, tissue exposure values (in roentgens) are reduced by moving the beam, but the total radiation (R-cm^2) into the body is not changed. This was illustrated in Figure 4-3.

RADIATION AND IMAGE QUALITY

One of the major compromises that must be made in imaging procedures using x-radiation is between patient exposure and image quality. Within certain limits, increasing image quality requires an increase in patient exposure. It is usually the specific image quality requirements of a procedure that determine the quantity of radiation that must be used in the imaging process. The three basic image quality factors (contrast sensitivity, detail, and noise) are each related to patient exposure. The film, screens, grid, and technique factors of an imaging procedure should be selected to produce adequate image quality with the lowest possible radiation exposure to the patient.

We now consider each factor that affects patient exposure and show how it relates to image quality.

FACTORS AFFECTING EXPOSURE

The exposure (dose) to a specific point within a patient's body is determined by a combination of factors. One of the most significant is whether the point in question is in or out of the primary beam. Points not located in the direct beam can receive exposure from scattered radiation, but this is generally much less than the exposure to points within the beam area. The factors that determine exposure levels to points within the body will be discussed in reference to the situation illustrated in Figure 23-3.

Receptor Sensitivity

One of the most significant factors is the amount of radiation that must be delivered to the receptor to form a useful image. This is determined by the sensitivity of the receptor. It was shown in Chapter 19 that there is a rather wide range of sensitivity values encountered in radiography. It is generally desirable to use the most sensitive receptor that will give adequate image quality. The exposure to points within the patient's body will be some multiple of the receptor exposure.

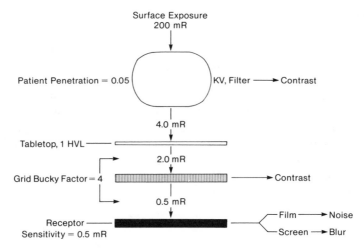

Figure 23-3 Factors that determine exposure values in radiography.

The sensitivity of a radiographic receptor is determined by characteristics of both the intensifying screen (Chapter 17) and the film (Chapter 14). To produce a net film density of 1, radiographic receptors require exposures ranging from 0.16 mR for 800-speed systems to more than 10 mR for some mammographic receptors. We will illustrate our immediate discussion using a receptor that requires a 0.5-mR exposure, as shown in Figure 23-3.

Intensifying Screens

The selection of intensifying screens for a specific procedure involves a compromise between exposure and image blur or detail. The screens that require the least exposure generally produce more image blur, as discussed in Chapter 17.

Films

Films with different sensitivity (speed) values are available for radiographic procedures. The primary disadvantage in using high-sensitivity film is that quantum noise is increased, as described in Chapter 19. In fact, it is possible to manufacture film that would require much less exposure than the film generally used. However, the image noise level would be unacceptable.

Grid

It was shown in Chapter 13 that the penetration of grids is generally in the range of 0.17 to 0.4 (17% to 40%). This corresponds to a Bucky factor ranging from 6 to 2.5. The exposure to the exit surface of the patient is the product of the receptor exposure and the grid Bucky factor. This is assuming that the receptor is not separated from the patient by a significant distance. The use of a high-ratio grid, which generally has a relatively low penetration, or high Bucky factor, tends to increase the ratio of patient-to-receptor exposure. Low-ratio grids reduce patient exposure by allowing more scattered radiation to contribute to the film exposure. In selecting grids the user should be aware of the general compromise between patient exposure and image contrast.

Tabletop

In many x-ray examinations, the receptor is located below the table surface that supports the patient's body. The attenuation of radiation by the tabletop increases the ratio of patient-to-receptor exposure. It is generally recommended that the tabletop have a penetration of at least 0.5 (not more than 1 HVL). The patient exposure with a tabletop that has a penetration of 0.5 will be double the exposure if no tabletop is located between the patient and the receptor.

Distance

Because of the diverging nature of an x-ray beam, the concentration of x-ray photons (exposure) decreases with distance from the focal spot. This is the inverse-square effect. This effect increases the ratio of patient-to-receptor exposure.

Consider a point located 20% of the way between the receptor surface and the focal spot. The magnification is 1.25. The exposure at this point is 1.56 times the receptor exposure because of the inverse-square effect. The distance between the surface, or point of interest, and receptor is generally fixed by the size of the patient. Therefore, the only factor that can be changed is the distance between the focal spot and the point of interest.

Patient exposure is reduced by using the greatest distance possible between the focal spot and body. The effect of decreasing this distance on patient exposure is illustrated in Figure 23-4; two body sections are shown with x-ray beams that cover the same receptor area. The x-ray beam with the shorter focal-to-patient distance covers a smaller area at the entrance surface. Because the same radiation

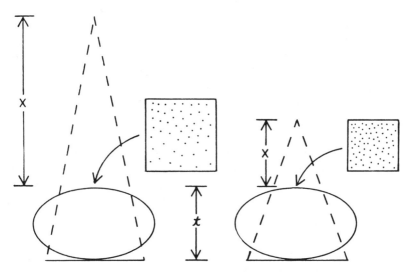

Figure 23-4 Decreasing the distance between the x-ray tube and the patient surface increases the concentration of radiation or surface exposure.

is concentrated into the smaller area, the exposure to the entrance surface and points within the patient is higher than with the greater focal-patient distance.

You should not locate the x-ray tube too close to the patient, for two reasons. It increases the exposure, as we have just discussed, and the image distortion will become more severe.

The inverse-square effect increases the concentration of radiation (exposure and dose) in the patient's body. However, the total amount of radiation (surface integral exposure) is not significantly increased by decreasing the tube-to-patient distance. The same radiation energy, or number of photons, is concentrated in a smaller area.

In procedures in which the body section is separated from the receptor surface to achieve magnification, exposure can significantly increase because of the inverse-square effect.

Tissue Penetration

If the point of interest (body organ) is not located at the exit surface of the body, the attenuation in the tissue layer between the organ and exit surface will further increase the exposure. The ratio of the organ-to-exit surface exposure is determined by the penetration of tissue.

The penetration of the tissue between the point of interest and the exit surface is determined by the distance between the two points, the type of tissue (lung, soft tissue, bone, etc), and the effective energy of the x-ray beam. For a given patient, the only factor that can be varied to alter penetration is the effective energy. This in turn depends on waveform, KV, and filtration. Generally speaking, three-phase or constant potential waveforms produce more penetrating radiation, which reduces patient exposure if all other factors are the same. It was shown earlier that adding filters to an x-ray beam selectively removes the low-energy, low-penetration photons. This produces an x-ray beam with a greater penetrating ability. Filtration of an x-ray beam is especially significant in reducing the exposure to points near the entrance surface. Patient exposure is generally reduced by increasing KV. The problem is that the higher KV values give lower image contrast because of object penetration and more scattered radiation.

Beam Limiting

Changing the x-ray beam area (or field of view) has relatively little effect on the entrance surface exposure but has a significant effect on the total amount of radiation delivered to the patient. The surface integral exposure is directly proportional to the beam area. A large beam will deliver more radiation to the body than a small beam if all other factors are equal.

Limiting the FOV to the smallest area that fulfills the clinical requirements is an effective method for reducing unnecessary patient exposure. Under no circumstances should an x-ray beam cover an area that is larger than the receptor.

Exposure Values

The entrance surface exposure for a radiographic procedure covers a considerable range because of variations in the factors discussed above. Table 23-1 gives some typical values for a variety of procedures.

Table 23-1 Typical Patient Exposure Values for Various X-Ray Procedures

Procedure	Exposure
Skull (L)	40 – 60 mR
Chest (L)	50 – 100 MR
Chest (PA)	10 – 30 mR
Breast	500 – 2,000 mR
Abdomen	100 – 400 mR
Lumbar Spine (L)	500 – 1,500 mR
Pelvis	250 – 500 mR
Fluoroscopy (1 min)	2,000 – 5,000 mR
Computed Tomography	1,000 – 4,000 mR

Study Activities

Explain the relationship between patient exposure and film sensitivity.

Explain why it is not always desirable to use intensifying screens with a high sensitivity to reduce patient exposure.

Explain how a grid affects patient exposure.

Describe how changing the FRD affects patient exposure.

Discuss the relationship between the field of view (image size) and patient exposure.

State the approximate surface exposure for the following procedures:

- PA chest
- skull
- abdomen
- three minutes of fluoroscopy

Chapter 24

Personnel Exposure and Protection

Personnel in the immediate vicinity of x-ray equipment or radioactive materials can be exposed to ionizing radiation. Therefore, certain actions must be taken to minimize their exposure and maintain it within acceptable levels.

This chapter covers the general concepts of radiation protection that apply to nonpatient personnel in a medical imaging facility.

EXPOSURE LIMITS

Since it is not practical to eliminate all human exposure, certain exposure limits have been established as part of radiation protection guidelines. The exposure limits established by the National Council on Radiation Protection (NCRP) are generally adopted by other agencies involved in radiation protection. The established exposure limits do not represent levels that ensure *absolute safety* but rather exposure levels that carry *acceptable risk* to the persons involved. The recommendations of the NCRP are in the form of *maximum permissible dose equivalent* (MPD) *values*. The limits are used in designing radiation facilities and in monitoring the effectiveness of safety practices.

The recommended MPD's vary with the occupational status of the individual and the parts of the body, as shown in Figure 24-1. The exposure limits are not for exposure received as a patient undergoing an x-ray examination.

Occupational Exposure

One set of exposure limits applies to persons (such as radiographers) receiving radiation exposure because of their work. This is *occupational exposure*. The MPD is 5 rem per year for all parts of the body that are not permitted higher limits.

261

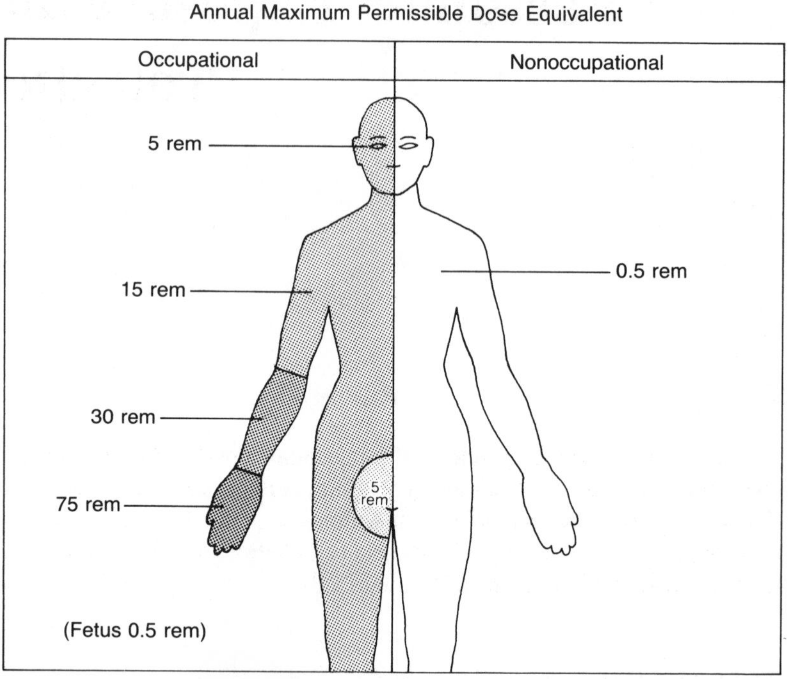

Annual Maximum Permissible Dose Equivalent

Figure 24-1 Maximum permissible dose equivalent (MPD) values.

The accumulated lifetime occupational exposure should not exceed 5 rem per year for each year past the age of 18.

The limbs and the skin have higher MPD values than the other parts of the body. The higher limit for skin is of little practical significance for x-radiation because it is virtually impossible to expose the skin without exposing the underlying tissue.

Nonoccupationally Exposed Persons

Persons who enter a facility as patients, visitors, or persons who do not routinely work in the facility might be exposed to radiation. The MPD for the nonoccupationally exposed person is 1/10th of the limit for occupationally exposed personnel.

Fetus

The MPD for a fetus is 0.5 rem for the total gestational period.

PROTECTION OBJECTIVES

The radiation safety practices in an x-ray facility are not designed to protect the personnel from all radiation exposure. It is not practical or even possible to always reduce exposure to zero. The efforts that would be required to completely eliminate all personnel exposure are not justified by the very small risk associated with the low levels of exposure that can be achieved. One general philosophy is to reduce personnel exposure As Low As Reasonably Achievable. This is known as the *ALARA principle*.

The MPD values described above are useful for monitoring the effectiveness of a radiation safety program. When personnel receive exposure that exceeds the MPD values, it usually indicates that something is not being done correctly. Exposure values that are less than the MPD are generally acceptable. However, if they can be reduced by a reasonable effort then it is advisable to do so.

The amount of exposure a person receives in an x-ray facility depends on many factors, including the design of the facility and the type of procedures being performed. However, a very important factor is the safety practices and habits of the individual radiographer.

SOURCE OF EXPOSURE

Patients receive exposure because they are in the primary x-ray beam. However, there is no legitimate reason for a radiographer to ever be in the direct beam when conducting a procedure. We receive our exposure from the radiation that is scattered from the patient's body, as shown in Figure 24-2.

The radiation that scatters from a patient's body is a very small fraction of the total x-ray energy going into the body. As a general rule of thumb, we usually assume that the scatter exposure (at a distance of 1 m from the center of the patient) is 0.1% of the entrance exposure to the patient. For example, if the surface entrance exposure to the patient is 1 R (1,000 mR), the scatter coming out of the side of the patient would be approximately 1 mR. This is only an approximation because the actual amount of scatter depends on several factors including the size of the patient and the field of view.

If someone is standing near a patient undergoing a typical x-ray procedure, the exposure from the scattered radiation would be just a few milliroentgens at most. The concern is when personnel work with many procedures, where the small exposures from each add up to a significant value.

Fluoroscopic procedures generally produce more scattered radiation than radiographic procedures because the total patient exposure is usually larger.

The total exposure received by personnel in an x-ray facility can be reduced by using a combination of three protective factors: (1) time, (2) distance, and (3) shielding. Let's now see how these factors can be applied.

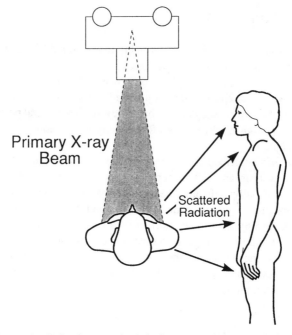

Figure 24-2 Scattered radiation from a patient's body.

TIME

Personnel exposure is accumulative. Therefore, if we reduce the time that one is close to the source of radiation exposure the total exposure will be reduced. The practical application of this is simple: Don't stay around a radiation source when it's not necessary.

DISTANCE

The exposure to an individual can be decreased by increasing the distance to the source of the radiation, the patient's body. This is because of the inverse-square effect that was introduced in Chapter 3. It also works for scattered radiation. According to this principle, the exposure will be reduced by a factor of 4 if you double the distance between yourself and the patient's body. For example, if you move from a distance of 5 feet to 10 feet away, you will receive only one fourth of the exposure.

The practical application of this principle is to stay as far away from the patient as possible during the actual exposure.

AREA SHIELDING

X-ray facilities are designed and built so that certain areas are shielded from the scattered radiation.

Adjacent Rooms

The walls of an x-ray room should be constructed so that personnel in adjacent (nearby) areas do not receive a radiation exposure that exceeds the MPD. If the walls are constructed of concrete or a similar building material, that often provides adequate shielding. If not, lead is added to the walls to reduce exposure to an acceptable level. The amount of shielding required in a specific wall should be determined by a certified radiological physicist, who will take into account factors such as the type of procedures conducted in the x-ray room and the occupancy of the nearby areas. The lead shielding is usually not visible because it is built into the wall.

Control Booth

Each radiographic room should have a control booth with sufficient shielding to protect the operator from most of the scattered radiation. Its window is made of a special glass that is a good x-ray absorber.

The x-ray equipment exposure switch should be located so that it cannot be operated from outside of the control booth protected area.

EQUIPMENT SHIELDING

There is a considerable amount of shielding built into the conventional fluoroscope. The housing for the image intensifier tube and the base on which it is mounted contain shielding material. This protects the fluoroscopist against both the primary x-ray beam and some of the scattered radiation coming from the patient's body. The lead drapes hanging from the intensifier tube base are an important source of shielding against scattered radiation.

The metal sides of the fluoroscopic table provide shielding against scattered radiation below the tabletop. This is the area of the most intense scattered radiation coming from the patient's body. Many tables are equipped with a slot through which the cassette is inserted into the Bucky mechanism. The cover to the slot should always be in place when the equipment is being used.

PERSONNEL SHIELDING

Whenever it is not possible to radiograph a patient from a protected area, a radiographer's body should be protected with appropriate shielding. This usually occurs when assisting with fluoroscopic procedures and when using portable radiographic equipment.

Lead Apron

A lead apron will provide adequate protection under most conditions. The amount of radiation a lead apron absorbs depends on the thickness of lead it contains. This is usually indicated by a small tag on the apron. Most aprons have a lead thickness that provides adequate protection under most conditions.

Lead aprons should be inspected periodically to determine if holes or tears have developed in the lead. This can be done by fluoroscoping them.

Aprons should be handled and stored properly to prevent damage to the lead.

Extremities

The levels of exposure encountered by a radiographer usually do not require shielding of the extremities. When physicians are palpating or performing interventional procedures in or near the fluoroscopic beam, the use of lead-lined gloves should be considered.

Head and Neck

Conventional radiographic and fluoroscopic procedures generally do not produce a significant exposure to the head and neck of the staff conducting the examination. However, there are some special procedures in which head and neck exposure must be considered. These are usually vascular examinations for which the x-ray equipment does not provide adequate shielding. The intensity of scattered radiation to the head and neck is also much greater when the x-ray tube is not positioned directly below the patient, as it is in a conventional fluoroscopic examination.

The two areas of concern are the thyroid gland and the eyes. Thyroid tissue is somewhat more sensitive to radiation-induced cancer than many other parts of the body. Even though this risk is relatively low, it increases with accumulated exposure to the thyroid gland. When individuals are working in conditions that produce a neck exposure exceeding the MPD, the use of a neck shield should be considered.

Cataracts can be produced when the accumulated exposure to the eye is on the order of several hundred roentgens. The eyes can be shielded by wearing glasses whose lenses are made of a radiation-absorbing material.

PERSONNEL MONITORING

Personnel who receive radiation exposure because of their occupation should be monitored and their accumulated exposures recorded in a permanent record. This is generally the responsibility of the employer. The exposure to individuals is determined by having them wear a personnel-monitoring device. Two types are often used: *film badges* and *thermoluminescent dosimeters* (TLD).

The general principle is to wear the badge at the body location that is most likely to receive the highest exposure in relationship to the MPD. This is generally on the body. If a lead apron is being worn, the collar (neck) area is usually the appropriate location. When personnel are conducting procedures with the hands very close to the primary x-ray beam, the use of wrist or ring badges should be considered.

Every x-ray facility has a designated *radiation safety officer* (RSO) who can answer specific questions about the use of monitoring devices.

Film Badge

The film badge is the most common personnel monitoring device used in x-ray facilities. The badge holder contains a small piece of film that is sensitive to radiation. The film is replaced periodically (usually monthly) and then processed in a special laboratory. The density of the processed film is measured to determine its accumulated exposure. A report of this exposure is then sent to the institution, which informs the individual and adds the report to the person's permanent file.

Thermoluminescent Dosimeter

A thermoluminescent dosimeter (TLD) is a device used for personnel monitoring in place of a film badge. Rather than film, it contains a small crystal of thermoluminescent material. When this material is exposed to radiation, a slight, invisible change occurs in the structure of the crystal. The amount of change is proportional to the accumulated exposure. The dosimeter crystal is collected periodically and placed in an instrument that can measure the amount of accumulated exposure. The measurement is made by heating the crystal. When the crystal is heated to a certain temperature, it emits light in proportion to the accumulated exposure. The amount of light coming from the crystal is measured by the instrument and converted to radiation exposure units. The word "thermoluminescence" refers to the process of heat (thermo) produced by light (luminescence).

The TLD can be used over and over because the effect of the radiation exposure is erased by each heating process.

Study Activities

State the maximum permissible dose limits for the various areas of your body.

State the maximum permissible dose limit for a visitor in an x-ray facility.

Identify the source of radiation that can expose personnel performing an x-ray procedure.

Describe the steps that you as a radiographer can take to reduce exposure to your body.

Index

269

size, framing and, 232, *232*
storage of, 136
subtraction, noise and, 205
Image blur. *See* Blurring
Image intensifier tube. *See* Intensifier tube
Image noise. *See* Noise
Inherent filtration, 107
Instantaneous value, in single-phase waveform, 82
Integral dose, *34, 35,* 36–37, *37*
Integral exposure, *29,* 34
Intensified radiography, image noise in, 202
Intensifier tube(s)
dual mode, *222*
fluoroscopic, 218–19
electronic gain in, *220*
function of, 220–21
gain characteristics, 219, *219*
minification, 219–20
steps in, 221
Intensifying screen(s)
blurring and, 177–78, *178*
conversion of x-ray energy in, *175*
description of, 173, *174*
film contact with, blurring and, 178, *179*
fluorescent material in, 173–74
function of, 19, 173
image detail and, 177–78, *178*
crossover, 178–79, *179, 180*
screen-film contact, 178, *179*
light emission by, 175
patient exposure and, 255
sensitivity of, 176–77
study activities on, 180
thickness of, blurring and, *178*
types of, 179
x-ray absorption by, 174–75
Intensity distribution, 187–88, *188*
Interaction(s)
Compton, 92–93, *93*
electron, 93–94, *94*
photoelectric, 92, *93*
attenuation in, 94–95, *95*
K edge in, *96,* 97

photon energy and, 95–97, *96*
photon-electron, 92, *93*
study activities on, 98
types of, 91, *92, 93*
of x-radiation with matter, 91
Interlacing, in fluoroscopy, 224
Inverse-square effect, 28–30, *29*
Inverse-square law, 30
Invertor(s), function of, 87
Iodine contrast, 117, *118*
Ion(s), 56–57
Ionization, by radiation electron, *94*
Ionizing radiation, 56

J

Joule (J)
in absorbed dose, 36
in x-ray tubes, 235–36

K

K edge, in photoelectric interactions, *96,* 97
keV. *See* Kiloelectron volt (keV)
Kiloelectron volt (keV), in photons, 27
Kilovolt peak (KV_p) values
in Bremsstrahlung process, 69, *70*
density control and, 210–11, *210*
x-ray penetration and, 106, *106*
minimum, *108*
Kilovoltage (KV)
in Bremsstrahlung process, 69, *70*
in characteristic radiation, 70
production of
autotransformer, *76,* 78–79
high-voltage transformer, *76,* 77–78
transformer principles, 76–77, *77*
by x-ray generator, 75–76
in x-ray production, 72–73, *72*
x-ray tube heating and, 236–37
Kilovoltage waveform, x-ray tube heating and, *242–43,* 246–48, *246, 247*
Kinetic energy, 43
in x-radiation, 66–67, *67*
KV. *See* Kilovoltage (KV)
KV_p. *See* Kilovolt peak (KV_p)